SpringerBriefs in Molecular Science

More information about this series at http://www.springer.com/series/8898

Xiaoming Sun · Liang Luo
Yun Kuang · Pengsong Li

Nanoseparation Using Density Gradient Ultracentrifugation

Mechanism, Methods and Applications

 Springer

Xiaoming Sun
State Key Laboratory of Chemical
    Resource Engineering
Beijing University of Chemical
    Technology
Beijing
China

Liang Luo
State Key Laboratory of Chemical
    Resource Engineering
Beijing University of Chemical
    Technology
Beijing
China

Yun Kuang
State Key Laboratory of Chemical
    Resource Engineering
Beijing University of Chemical
    Technology
Beijing
China

Pengsong Li
State Key Laboratory of Chemical
    Resource Engineering
Beijing University of Chemical
    Technology
Beijing
China

ISSN 2191-5407   ISSN 2191-5415   (electronic)
SpringerBriefs in Molecular Science
ISBN 978-981-10-5189-0   ISBN 978-981-10-5190-6   (eBook)
https://doi.org/10.1007/978-981-10-5190-6

Library of Congress Control Number: 2017959543

Printed on acid-free paper

This Springer imprint is published by the registered company Springer Nature Singapore Pte Ltd.
The registered company address is: 152 Beach Road, #21-01/04 Gateway East, Singapore 189721, Singapore

Preface

Since the discovery of unique properties of nanomaterials, colloidal nanoparticles (NPs) with tunable size and shape have attracted vast attention due to their wide applications in catalysis, energy conversion, bioissues, etc. Therefore, size and geometric control of nanomaterials are important to the discovery of intrinsic size-/shape-dependent properties and bottom-up approaches for the fabrication of functional nanodevices. Nowadays, two general strategies have been employed to create size-uniform nanostructures. One method is direct particle size control during synthesis by adjusting growth parameters; however, it usually needs very harsh synthetic conditions to achieve relatively uniform size distributions, this strategy only suits to quite limited systems, and the parameters must be strictly followed, otherwise, in most cases, the products are only roughly "monodisperse" with certain deviations.

As effective complementary process to synthesis optimization, nanoseparation is attracting more and more interest for providing strictly monodisperse NPs. A variety of separation methods, including selective precipitation, magnetic separation, filtration/diafiltration, electrophoresis and chromatographic methods, have been explored as different ways of attaining particle fractions with narrow shape and size distributions. The density gradient ultracentrifugation (DGUC) method, as a general, nondestructive and scalable separation method adapted from biomacromolecular separation technology, has recently demonstrated as a versatile method for acquisition of monodispersed colloidal NPs which are hard to be synthesized. This separation method was applicable to both aqueous (polar) and organic (non-polar) solvents system, and NPs with different size, density and morphology can be separated. Separation objects involve nearly all kinds of materials including metal and metal oxides/sulphides, carbon materials, semiconductors. And the separation based on cluster sizes is also demonstrated. Based on the careful characterization on fractions with "purer colloids", some key parameters controlling the growth or phase conversion are uncovered.

In this book, the classification and mechanism of DGUC will be introduced, and various separation examples will be discussed to show the versatility of such an efficient separation technique. Synthesis–structure–property relationships would be observed on the separated NPs, which could even guide synthetic optimization. Besides basic DGUC separation, concentration and purification of NPs could be achieved at the same time when water/oil interfaces were introduced into the separation system. Furthermore, by introducing a reaction zone or an assembly zone in the gradient, we can even find the surface reaction and assembly mechanisms of NPs since reaction time could be precisely controlled and chemical environments could be changed extremely fast. Finally, a strict mathematical description and computational optimization model would be given to predict the best separation parameters for a given colloidal system, making the DGUC method an efficient, practical and predictable separation method.

In short, DGUC-based "lab in a tube" method would not only provide an efficient separation tool for various nanostructures, but also pave a way for the researches on synthesis optimization, assembly and surface reactions, which laid the cornerstone of the development of nanotechnology.

Beijing, China   Xiaoming Sun
                                                                      Yun Kuang

Contents

Chapter 1
Introduction to Nanoseparation

Yun Kuang, Ming Jiang and Kai Sun

Abstract Nanomaterials have been attracted tremendous attentions for decades, due to their unique properties on nanoscale. As well known, the properties, such as chemical, thermal, mechanical, optical, electrical, and magnetic properties, are highly dependent on the size of nanomaterials, as so-called size-dependent quantum effect. Thus, to obtain monodisperse nanostructures is of great significance. With the help of various ligands, solution-phase synthesis could produce colloidal nanostructures with relatively homogeneous morphology and narrow size distribution for some nanosystems. However, owing to the synthetic difficulties, fine control of uniform nanostructures still remains a big challenge. Besides, nanoseparation, as a "post-synthesis" method, is a powerful tool to sort and achieve monodispersity and to avoid possible aggregation of the colloids. In this chapter, the basic principles of nanoseparation and a brief introduction of common techniques used for the separation of nanostructures, including membrane filtration, chromatograph, electrophoresis, magnetic field and centrifugation, will be discussed.

Keywords Size-dependent effect · Monodispersity · Post-synthesis
Principles of nanoseparation · Common techniques for nanoseparation

1.1 Challenges for Nanomaterial Synthesis—Monodispersity

The concept of "nano" was first discussed in 1959 by renowned physicist Richard Feynman in his talk "There's Plenty of Room at the Bottom," in which he described the possibility of synthesis via direct manipulation of atoms. For the past half-century, the fast development of nanoscience and nanotechnology has brought scientists new sights into such an unusual scale, in which some new phenomena and materials are springing up.

One nanometer (nm) is one billionth, or 10^{-9}, of a meter. The commonly called nanomaterial is a kind of material with at least one-dimensional size in the range of 1–100 nm. When the size of solid matter is comparable to what can be seen in a regular

© The Author(s) 2018
X. Sun et al., *Nanoseparation Using Density Gradient Ultracentrifugation*,
SpringerBriefs in Molecular Science, https://doi.org/10.1007/978-981-10-5190-6_1

optical microscope, there is little difference in the properties of the particles. While the size of materials is as low as 1–100 nm, various properties, such as chemical, thermal, mechanical, optical, electrical, magnetic properties, show obvious size-dependent phenomena and are significantly different from those at larger scales. This is the size scale where so-called quantum effects rule the behaviors and properties of the materials. Thus, scientists working in the areas of physics, chemistry, material science, and engineering have devoted great efforts into size control of materials in nanoscale within the past 30 years, to fabricate monodisperse nanostructures. Generally, there are two strategies for the fabrication of nanomaterials: One is top-down method, but due to the limits of current facilities, the size control of nanomaterials can only be achieved in one or two dimensions, and the size is usually above 50 nm with accuracy larger than 10 nm. The other way is bottom-up method. With the help of various ligands, solution-phase synthesis could produce nanostructures with relatively homogeneous morphology and narrow size distribution. However, because of non-uniform temperature field and local concentration field, size deviation would inevitably exist in one-pot synthesis, and thus, there are always some byproducts in wet synthesis. While for chemical vapor synthesis or solid synthesis, such non-uniformities are more obvious, and the final products are always polydispersed.

Therefore, for practical use of nanomaterials with specific properties, how to obtain monodisperse nanostructures is an urgent problem. One solution is direct size control during synthesis by applying very harsh synthetic conditions, but this strategy only suits quite limited systems, and the parameters must be strictly controlled; otherwise, in most cases, the products are only roughly "monodisperse" with certain deviations. The other way is to develop a new separation and purification method to get strict monodisperse nanomaterials.

1.2 Common Techniques Used for Separation of Nanostructures

Employing suitable techniques for the separation of nanomaterials is necessary to achieve monodispersity and to avoid possible aggregation of the colloids. Thus, finding separation methods of colloidal nanoparticles with high efficiency is one of the main challenges in nanomaterials research.

Typically, an ideal separation method should follow six principles: (1) high efficiency to separate one sample into as many as fractions with obvious difference; (2) high versatility toward different systems, for instance, 2D nanosheets, 1D nanorods/nanowires, and 0D nanodots; (3) wide range of feasibility for nanoparticles with wide size ranges (e.g., 1–500 nm) and colloidal nanoparticle systems with different solubility; (4) little sample loss. Try to avoid any possible "solid–solid" (sample solid separation media) contact to minimize sample loss, or efficiency loss of the separation system; (5) the intrinsic properties of colloidal particles after separation should not be changed; even their ligands should be well protected while any possible aggregation of the colloids should be avoided; (6) the system for separation is reusable or easy to duplicate and/or optimize. By parameters optimization in secondary (or repeated) separation, the separation efficiency can be further improved. Currently, a variety of separation techniques, including electrophoresis, chromatography, magnetic field, membrane filtration, and centrifugation, have been used to separate different sized materials. Among all, density gradient ultracentrifugation (DGUC) emerges due to its advantage of high efficiency and low cost beyond commonly used differential centrifugation and attracts wide attentions. In this section, a general introduction of the main separation techniques will be given and the following chapters will mainly focus on the detailed description of DGUC.

1.2.1 Membrane Flirtation

Membrane filtration emerged as a promising separation technique in the early twentieth century and rose rapidly in the 1960s. This technique is based on porous membrane wherein pore size dictates the retention and elution of materials from samples, i.e., the cutoff size [1]. On the basis of pore size, membrane can be divided into reverse osmosis membrane, microfiltration membrane, ultrafiltration membrane, and nanofiltration membrane, while on the basis of polarity of membrane materials it can be further divided into organic and inorganic membranes. Inorganic membrane mainly consisted of microfiltration membrane, including ceramic membrane and metallic membrane. The pores are made mainly by pyrolysis of the template in ceramic case and by dealloying in metal case, while organic membrane is made of polymer materials, which can meet the requirements of nanofiltration. The primary components of organic membranes are polyether sulfone, aromatic polyamide, cellulose acetate, poly-fluoropolymer, and so on. The pores are made mainly by sputtering, interfacial polymerization, and so on.

Membrane filtration has been used widely for water purification or wastewater treatment and recently extended to the separation of biological materials and colloidal nanostructures due to its affordability, convenience, and versatility. This approach has been introduced into the separation and purification of nanosized samples along with the development of technology. Generally, membrane filtration only requires a simple pumping filtration operation, but it suffers from some disadvantages such as blocking, unsustainable, and small output.

James E. Hutchison et al. successfully separated and purified the water-soluble gold nanoparticles with diameters distribution ranging from 1.5 to 3 nm by continuous diafiltration and demonstrated that the higher purity afforded by diafiltration can allow for more precise determination of electronic and optical properties of gold nanoparticles [2]. As shown in Fig. 1.1, small molecule impurities or small nanoparticles penetrated through the pore on the membrane during the continuous flowing process of separating medium and were separated from original solution eventually.

Prabha and Vinod Labhasetwar utilized double emulsion solvent evaporation technique with nanofiltration membrane to separate PLGA (poly (D,L-lactide-co-

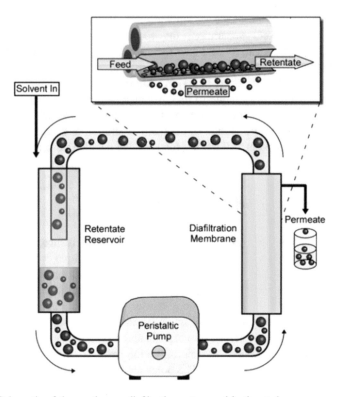

Fig. 1.1 Schematic of the continuous diafiltration setup used in the study

glycolide)) nanoparticles contained DNA molecules and finally gave a heterogeneous particle size distribution [3]. Figure 1.2 clearly shows the contract: (a) particle size distribution of unfractionated nanoparticles; (b) nanoparticles passing through the membrane; (c) nanoparticles retained on the membrane. Furthermore, they found that the smaller-sized fraction of nanoparticles is with significantly higher transfection efficiency as compared to the larger-sized fraction.

Anne M. Mayer et al. introduced a novel thin film composite nanofiltration membrane into the separation and purification of monomolecular layer-protected gold nanoparticles and concluded that on the basis of composite nanofiltration membrane the selectivity and flux performance of nanofiltration method can be further improved [4].

1.2.2   Chromatography

Chromatography technique has been widely used in analytical chemistry for the separation of components from a mixture based on their differences in physical

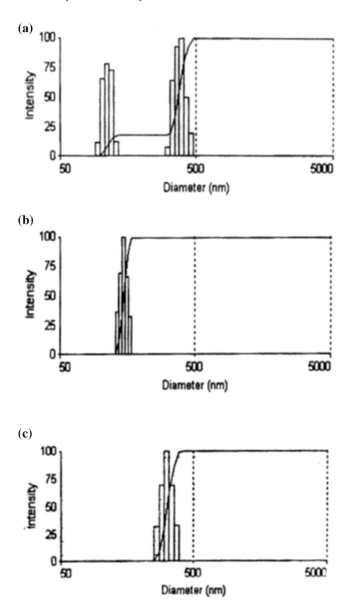

Fig. 1.2 Particle size distribution of **a** unfractionated nanoparticles, **b** nanoparticles passing through the membrane, and **c** nanoparticles retained on the membrane

adsorption or distribution properties. Gas or column chromatography is more suitable for the separation of small molecules. While for colloidal nanoparticles, which can be regarded as "huge inorganic molecules" judging from the size and molecule weights, those chromatography methods used for the separation of macromolecules turn out as optional choices. For instance, ion chromatography (IC) and size exclusion chromatography (SEC) techniques can be used for the separation of charged nanoparticles or different sized nanoparticles. In ion chromatography (IC), ions and polar molecules can be separated on the basis of their ion-exchange affinity [5]. Size exclusion chromatography (SEC) [6, 7] is also known as gel permeation chromatography (GPC), which is a powerful tool for probing the size distribution of complex materials.

Stationary phases used in SEC usually consist of chemically inert and porous gels with pore diameter approximately ranging from 1 to 100 nm, such as polyacrylamide gel, Sephadex gel, Sepharose gel, and polystyrene gel. The separation mechanism of SEC depends on the size and volume distribution of the products, rather than the interacting forces between the materials and gels. When the nanoparticles flow through the column along with the mobile phase, the nanoparticles with large volumes will directly outflow the column in advance because they cannot penetrate into the pores of the gel, while the materials with medium size are able to enter into the pores partially and the smallest particles can fully penetrate into the gel. Consequently, the materials get varied retention time and thus can be separated by size/volume under appropriate control. Elution behavior of SEC is generally influenced by the pore size distribution, types of mobile phases, and fluid dynamics of materials.

Fischer and co-workers successfully accomplished the separation of water-soluble CdS nanoparticles through chromatography in 1989 for the first time and gave a distinct size distribution of nanoparticles [8]. After that, organic-phase-soluble CdS nanoparticles were also separated via the same technique by Krueger et al. [6]. Figure 1.3 illustrates the separation. Clearly, the larger tetrapods eluted earlier than the spherical dots.

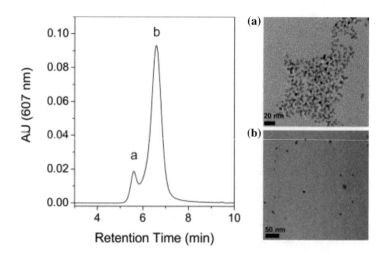

Fig. 1.3 Single-pass separation of polydisperse sample into multipods and quantum dots. TEM was performed on collected fractions of peaks (**a**) and (**b**)

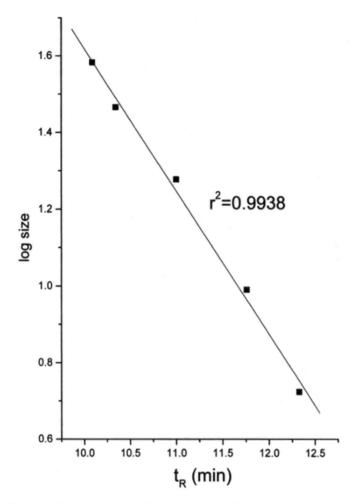

Fig. 1.4 Semi-logarithmic plot of the diameter as a function of the retention time for gold nanoparticles in SEC. Sample volume: 10-μl gold particles solution; mobile phase: 5-mM SDS; flow-rate: 0.5 ml/min

On the basis of Fischer's work, Vicki L. Colvin et al. demonstrated that SEC method can also be applied as a promising method for standardizing nanocrystalline materials directly in the solution state [6]. Moreover, Guor-Tzo Wei et al. applied SEC method in the separation of gold nanoparticles, and an excellent linear relationship for the logarithm of the particle size ranges from 5.3 to 38.3 nm as a function of elution time was found [9]. The linear relationship was explained in Fig. 1.4, which strongly implied that the separation mechanism in this work was based on steric exclusion. Royce W. Murray et al. further separated gold nanoclusters with protective monolayers by SEC [10]. Benjamin S. Flavel et al. demonstrated the separation of double-walled carbon nanotubes (DWCNTs) from single-walled carbon nanotubes (SWCNTs) containing starting material, using fast, easily scalable, and financially viable SEC which provided a convenient avenue to prepare enriched DWCNTs in a straightforward and easily scalable manner [11].

The DNA/CNT hybrid, formed by wrapping single-stranded DNA around single-walled carbon nanotubes, which are both stable and dispersible in aqueous solution, can also be separated by ion-exchange chromatography. The hybrid elutes at an ionic strength that depends on the electronic character and diameter of the core nanotube. Thus, this work provides a systematic mechanism for separating the nanotubes by using chirality. Ming Zheng and co-workers separated nanotubes and presented a theoretical model for this separation process that explained mainly all the salient features observed experimentally, and provided accurate predictions for critical elution salt concentration [12].

The main advantage of chromatography technique is that the separation process can be rationally designed, and the compositions of the stationary phase and mobile phase can be turned to match with the nature of colloidal particles. However, the quality of separated production can only reach in milligram level and the separated nanoparticles might get stuck in the pores of stationary phase in SEC case or adhere on the surface in the IEC case, and be hard to elute, which could decrease the reproducibility and increase the cost.

1.2.3 Electrophoresis

The separation mechanism of electrophoresis relies upon the surface charges of nanostructured materials, which drive charged particulates move toward the electrode of opposite charge under the effect of external electric fields [13, 14]. The separation efficiency is strongly influenced by many factors, such as types of buffer solutions, pH value, concentration of solutions, ionic strength, and the electric field intensity. To obtain a more quantitative analysis, Eq. 1.1 is employed to express the electrophoretic mobility of ionic species [15]:

$$\mu_p = \frac{q}{6\pi\eta r} \tag{1.1}$$

where q is the charge of the ion or colloidal particle, η is the coefficient of viscosity of the fluid, and r is the hydrodynamic radius of the ion. The principle of electrophoretic separation is endowed with wide applications, which gives rise to the further development in a variety of techniques for electrophoretic separation, such as capillary electrophoresis (CE) [16], gel electrophoresis (GE) [17, 18], electric field flow fractionation (EFFF) [19, 20], and isoelectric focusing technique (IFT) [20].

Gel electrophoresis (GE) is a technique that is applying a gel (polyacrylamide gel, Sepharose gel, etc.) as supporting material to accomplish a successful separation. The first application of GE in preparation of monodisperse Q-sized particles was given by H. Weller back in 1989 [5]. His group obtained highly monodisperse semiconductor colloids by employing GE, which is a well-known technique in biochemistry for the separation of biomolecules and CdS. Applying the same technique, Robert L. Whetten further separated Au clusters clad with monolayer glutathione by using polyacrylamide gel [16]. Walter A. Scrivens and co-workers demonstrated the efficiency of preparative electrophoresis using agarose gels with approximately 100-nm pore size to separate Au nanoparticles based on size and shape [13]. Furthermore, Carsten So lnnichsen's group also employed GE to achieve a successful separation of Ag nanoparticles [18]. The highly enriched nanosheets, nanorods, and nanoparticles of Ag were obtained, and the strong influence of size and shape on the frequency or wavelength of the plasmon resonance was also investigated.

Capillary electrophoresis (CE) is a separation technique that utilizes capillary zone as the channel to accomplish the separation of nanoparticles under the external electric field. Because of the capillary zone, the separation of CE can be achieved by the mobility of the species depending on the solvent medium, and the charge, size, and shape of nanoparticles. Fu-Ken Liu's group separated Au/Ag core/shell NPs with the diameter range from 25 to 90 nm by using CE [21]. After that, Doo Wan Boo et al. demonstrated the application of CE in the separation of polystyrene nanospheres and Au nanoparticles, verifying the advantages of CE in size- and shape-dependent separation [15]. However, the biggest obstacle in this technique is the extremely high applied voltage power and very limited yield.

Though the continuous free-flow electrophoresis (CFE) also needs external electric field as driving force for separation, the applied voltage is much lower than that of CE method and the production can reach to several grams, exhibiting high efficiency. David E. Clittel and co-workers investigated the CFE as a technique to isolate the monodisperse samples of water-soluble tiopronin Au monolayer-protected nanoclusters (MPCs) [19]. Figure 1.5 shows the schematic of the CFE fractionation of water-soluble monolayer-protected nanoclusters.

Matthias Hanauer and co-workers demonstrated the separation of gold and silver nanoparticles according to their size and shape by agarose gel electrophoresis after coating charged polymer layer [18]. As shown in Fig. 1.6a, the sample was the mixture of spheres, triangles, and rods. After the running for 30 min at 150 V in $0.5\times$ TBE buffer (pH $\approx$ 9), the fractions were successfully separated, which was verified by the isolated colors distribution due to the size- and shape-dependent optical properties of gold and silver particles. The detailed statistics of the qualitative sample composition revealed by TEM images (Fig. 1.7) further confirm the separation efficiency.

Subsequently, I-Ming Hsing's group developed the electric field flow fractionation (EFFF) for nanoseparation, which is conducted by changing the electric field from constant voltage operation to a pulsed voltage operation [22]. Lespes et al. employed cyclical electrical field flow fractionation (CyEIFFF) for the separation of Au nanoparticles and carbon nanotubes [23]. They found that the higher electric field used in CyEIFFF is potentially more powerful and can be applied to maintain charged molecules through electrophoretic mobility (μ) differences.

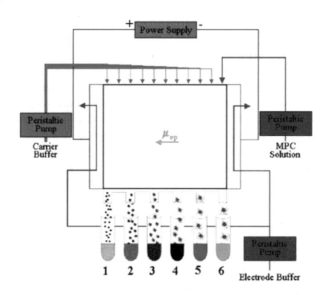

Fig. 1.5 Schematic of the CFE fractionation of water-soluble monolayer-protected nanoclusters

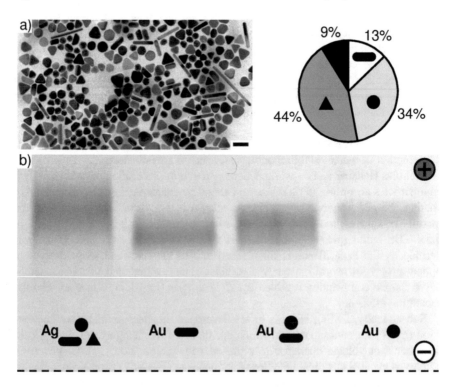

Fig. 1.6 a Typical TEM picture of a silver nanoparticle sample and the proportion of spheres, triangles, and rods. **b** True color photograph of a 0.2% agarose gel run for 30 min at 150 V in 0.5× TBE buffer (pH ≈ 9)

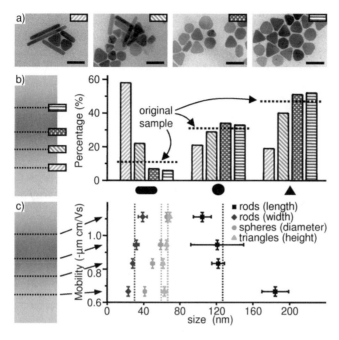

Fig. 1.7 a Typical TEM picture; **b** the tendency of the result by detailed statistics of the qualitative sample composition in terms of rods, spheres, and triangles; **c** is the careful quantitative analysis which shows a separation of particles according to sphere diameter and rod aspect ratio

Fig. 1.8 Process of IEF for the separation of the colloidal particles

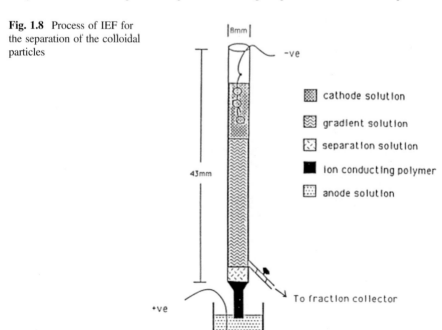

Isoelectric focusing (IEF) is commonly used to determine the isoelectric point of proteins and enzymes. Murali Sastry et al. introduced IEF into the separation for nanoparticles and successfully accomplished the separation of Au and Ag nanoparticles which were capped with 4-carboxythiophenol [20]. Figure 1.8 shows the typical process of IEF for the separation of the colloidal particles. A gradient solution affected by pH can be found in the middle of the column. The nanoparticles capped with 4-carboxythiophenol can reach to pH at which it is electrically neutral under the influence of the electrostatic field and finished the separation process efficiently.

Electrophoresis has become one of the most useful tools in separation science because of its high separation efficiency, low cost, versatility, ease of sample preparation, and automation. However, some limitations of electrophoresis such as poor concentration, sensitivity due to its lower sample loading, and shorter optical path length limit its further applications in separation science.

1.2.4 Magnetic Field

Magnetic method is developed for the separation of magnetic fluids depending on the magnetic properties of different materials [24]. Typically, a common magnetic fluid is fractionated with different washing solutions and the fractions are isolated by their size and magnetic properties. Figure 1.9 shows the retention rates of different sized particles, which determine their magnetic behaviors [25].

Thomas Rheinla Knder et al. presented the fractionation of magnetic nanoparticles by the simple magnetic method and demonstrated the high separation capacity of magnetic method and without addition of electrolyte as compared with other non-magnetic fractionation techniques [25].

Based on the same conception, Mary Elizabeth Williams and co-workers developed capillary magnetic field flow fractionation (MFFF) method for the purification and analysis of magnetic nanoparticles [24]. Figure 1.10 shows the systematic separation mechanism of the MFFF method in which the particles that interact weakly with the magnetic field eluted with the shorter period of retention time while strongly interacted particles with magnetic field eluted with the longer period of retention time. The samples of magnetic nanoparticles composed of either γ-Fe_2O_3 (maghemite) or $CoFe_2O_4$ with average diameters ranging from 4 to 13 nm were separated to validate the practicability of this method. As shown in Fig. 1.11, the excellent consequence declares that further development and automation of

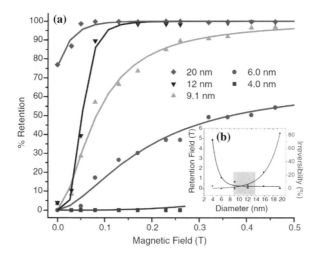

Fig. 1.9 Retention rates of particles with different sizes

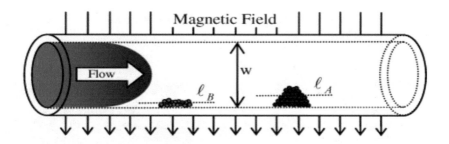

Fig. 1.10 Diagram of the MFFF separation mechanism

MFFF may provide a facile way to purify polydisperse samples for use in a wide variety of biomedical applications.

Mary Elizabeth Williams and co-workers successfully developed differential magnetic catch and release (DMCR) method for the separation of $CoFe_2O_4$ particles. In addition, they also used the DMCR method to separate and purify hybrid $Au–Fe_3O_4$ and $FePt–Fe_3O_4$ nanoparticles, and the magnetism of nanoparticles was also characterized by a magnetometer [26].

Vicki L. Colvin and co-workers developed the method to obtain monodisperse Fe_3O_4 nanocrystals with very low magnetic field gradient [27] (<100 T/m) (Fig. 1.12). High surface area and monodisperse magnetite (Fe_3O_4) nanocrystals

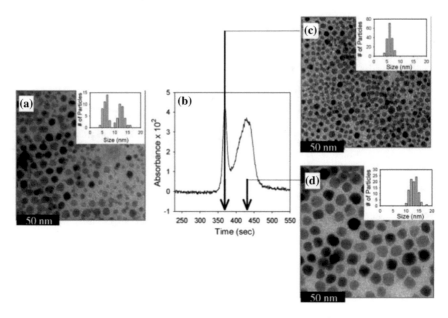

Fig. 1.11 TEM images of the nanoparticles before (**a**) and after (**c, d**) the MFFF separation, **b** UV-vis spectrum of mixed nanoparticles

Fig. 1.12 Magnetic batch separation of 16-nm water-soluble Fe_3O_4 NCs with a conventional separator (Dexter Magnetic LifeSep 50SX)

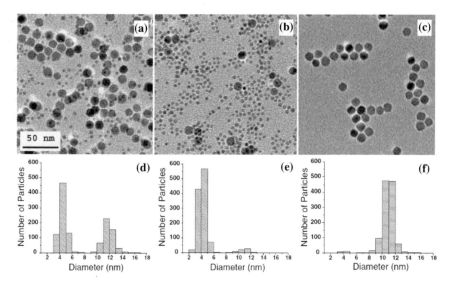

Fig. 1.13 a Transmission electron microscopic (TEM) micrograph of the initial bimodal mixture. **b** TEM micrograph of the high-field (0.3 T) fraction. **c** TEM micrograph of low-field (0.03 T) fraction. (**d–f**) Size distribution histograms for particles counted in (**a–c**), respectively

(NCs) are distributed in a size-dependent fashion. Further, mixture of 4.0-nm and 12-nm Fe_3O_4 nanoparticles with a 1:3 ratio (v/v) was chosen to verify the separation efficiency, and as revealed by the qualitative statistics based on TEM images (Fig. 1.13), Fe_3O_4 nanoparticles were successfully separated after the treatment of different intensities of magnetic field.

Though magnetic method is a powerful tool due to its high efficiency and convenience, it can only be used for the separation of magnetic materials.

1.2.5   Centrifugation

Centrifugation is a process which involves the utilization of the different centripetal forces for the sedimentation of heterogeneous mixtures [28]. Driven by the centrifugal force during the centrifugal process, the suspended solids precipitate from the solution, and compositions with density variation can be separated stepwise due to their difference in the rate of sedimentation. Owing to the unique characteristic, centrifugation has been widely applied in many areas, such as colloidal chemistry, polymer chemistry, biological chemistry, and environmental conservation. Centrifugation has been developed as a variety of forms along with the development of equipment. Differential centrifugation and density gradient ultracentrifugation (DGUC) are two main component sections of centrifugation, and DGUC can be further divided into rate-zonal centrifugation and density gradient centrifugation (Fig. 1.14). The classification of centrifugation is demonstrated below:

Centrifugation has been applied for over a decade to separate and purify nanoparticles. To improve the size and morphology distribution of products, ordinary centrifugation is routinely used to achieve efficient separation. Guo et al. utilized the shape-relative stability differences of gold nanorods/nanosheets to realize the salt-triggered separation of colloidal anisotropic gold nanoparticles, as shown in Fig. 1.15 [29].

Vivek Sharma et al. analyzed different hydrodynamics of gold nanorods and nanospheres and successfully accomplished the shape separation of nanoparticles. The related process is demonstrated (Fig. 1.16) [30],

Vicki L. Colvin and co-workers accomplished that the separation of different nanoparticles relies on sedimentation properties of various material [31]. The empirical sedimentation property, S_a, was defined over 80 years ago by Svedberg [32]:

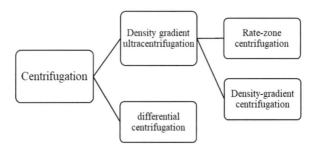

Fig. 1.14 Classification of centrifugation

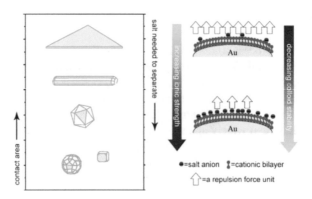

Fig. 1.15 Schematic of salt-triggered separation of anisotropic gold NPs from co-produced isotropic NPs based on shape-relative solution stability differences

(a)   (b)

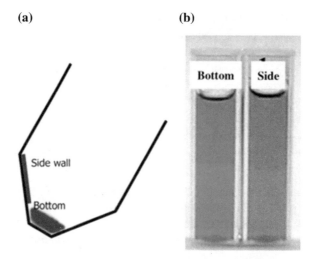

Fig. 1.16 a Distribution of separated Au nanoparticles in centrifuge tube. **b** Camera images of separated Au nanoparticles

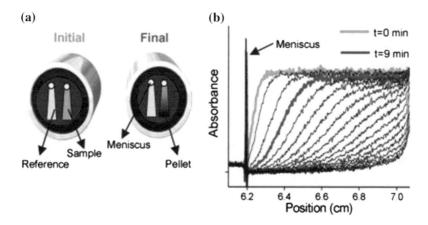

Fig. 1.17 Overview of the analytical ultracentrifugation process

$$S_a = \ln\left(\frac{r_b}{r_m}\right)\left[\omega^2(t_b - t_m)\right]^{-1} \tag{1.2}$$

where ω is the angular velocity, and t_b and t_m are the time in seconds, at positions r_b and r_m, respectively.

Typical centrifugation process is shown in Fig. 1.17. There is no change in reference centrifuge tube after centrifugation, while nanocrystals would sediment and ultimately form a pellet at the bottom of the centrifuge tube. The time-dependent

Table 1.1 Evaluation of common separation techniques (5 points means the best)

Separation technique	Separation efficiency	Feasibility	Cost	Throughput	Reusability	Sample loss	Average
Membrane filtration	4	4	2	5	2	5	3.67
Chromatography	5	3	3	2	5	3	3.50
Electrophoresis	3	2	3	3	2	3	2.67
Magnetic field	3	2	3	3	4	3	3.00
Differential centrifugation	3	3	5	3	5	3	3.67
Ultracentrifugation	5	5	4	4	5	3	4.33

absorbance variation of fractions along the centrifuge tube further confirms the sedimentation process [31]. Based on this principle, they found appropriate sedimentation coefficients for setting the optimal centrifugation rates for the separation of mixtures and successfully separated Fe_3O_4 from Au nanoparticles.

Density gradient ultracentrifugation (DGUC) has been successfully applied in the separation and analysis of various nanoscaled materials. As a general, nondestructive, and scalable separation method, DGUC was recently been demonstrated as an efficient way of sorting colloidal nanoparticles according to their differences in chemical, structural, size, or morphology [33]. Developing specific mathematical optimization model and separation parameters of DGUC technique toward different systems can be efficient, practical, and predictable for the broader applications. We will give the comprehensive and detailed introduction of DGUC in the succeeding sections.

As mentioned above, numerous techniques have been widely investigated to separate the nanomaterials from different solutions. These common separation techniques are demonstrated in Table 1.1 with brief summary. Though these analytical methods have been successfully applied in many systems, the limitations of these techniques restrict their comprehensive application. Enhanced method with high precision and efficiency should be introduced in the separation of nanostructures.

References

1. Ersahin ME, Ozgun H, Dereli RK et al (2012) A review on dynamic membrane filtration: materials, applications and future perspectives. Bioresource Technol 122(5):196–206
2. Sweeney SF, Woehrle GH, Hutchison JE (2006) J Am Chem Soc 128:3190–3197
3. Prabha S, Zhou WZ, Panyam J et al (2002) Size-dependency of nanoparticle-mediated gene transfection: studies with fractionated nanoparticles. Int J Pharm 244:105–115
4. Akthakul A, Hochbaum AI, Stellacci F et al (2005) Mayes size fractionation of metal nanoparticles by membrane filtration. Adv Mater 17:532–535
5. Weiss J (2008) Ion chromatography. Wiley, Hoboken
6. Krueger KM, Al-Somali AM, Falkner JC et al (2005) Colvin characterization of nanocrystalline CdSe by size exclusion chromatography. Anal Chem 77:3511–3515

7. Novak JP, Nickerson C, Franzen S et al (2001) Purification of molecularly bridged metal nanoparticle arrays by centrifugation and size exclusion chromatography. Anal Chem 73:5758–5761
8. Fischer CH, Weller H, Katsikas L et al (2002) Photochemistry of colloidal semiconductors. 30. HPLC investigation of small CdS particles. Langmuir 5(2):429–432
9. Wei GT, Liu FK, Wang CRC (1999) Shape separation of nanometer gold particles by size-exclusion chromatography. Anal Chem 71(11):2085–2091
10. Jimenez VL, Leopold MC, Mazzitelli C et al (2003) HPLC of monolayer-protected gold nanoclusters. Anal Chem 75(2):199–206
11. Moore KE, Pfohl M, Hennrich F et al (2014) Separation of double-walled carbon nanotubes by size exclusion column chromatography. ACS Nano 8(7):6756–6764
12. Anand Jagota SRL, Constantine Khripin, Ming Z (2005) Theory of structure-based carbon nanotube separations by ion-exchange chromatography of DNA/CNT hybrids. J Phys Chem B 109(7):2559–2566
13. Xu X, Caswell KK, Tucker E et al (2007) Size and shape separation of gold nanoparticles with preparative gel electrophoresis. J Chromatogr A 1167(1):35–41
14. Arnaud I, Abid JP, Roussel C et al (2005) Size-selective separation of gold nanoparticles using isoelectric focusing electrophoresis. Chem Commun 6(6):787
15. Hwang WM, Lee CY, Boo DW et al (2003) Separation of nanoparticles in different sizes and compositions by capillary electrophoresis. B Kor Chem Soc 24(5):684–686
16. Schaaff TG, Knight G, Shafigullin MN et al (1998) Isolation and selected properties of a 10.4 kDa gold:glutathione cluster compound. J Phys Chem B 102(52)
17. Eychmüller A, Katsikas L, Weller H (1990) Photochemistry of semiconductor colloids. 35. Size separation of colloidal cadmium sulfide by gel electrophoresis. Langmuir 6(10):1605–1608
18. Hanauer M, Pierrat S, Zins I et al (2007) Separation of nanoparticles by gel electrophoresis according to size and shape. Nano Lett 7(7):2881–2885
19. Peterson RR, Cliffel DE (2005) Continuous free-flow electrophoresis of water-soluble monolayer-protected clusters. Anal Chem 77(14):4348–4353
20. Gole AM, Sathivel C, Lachke A et al (1999) Size separation of colloidal nanoparticles using a miniscale isoelectric focusing technique. J Chromatogr A 848(1–2):485–490
21. Liu FK, Tsai MH, Hsu YC et al (2006) Analytical separation of Au/Ag core/shell nanoparticles by capillary electrophoresis. J Chromatogr A 1133(1):340–346
22. Lao AIK, Trau D, Hsing IM (2002) Miniaturized flow fractionation device assisted by a pulsed electric field for nanoparticle separation. Anal Chem 74(20):5364
23. Gigault J, Gale BK, Le Hecho I et al (2011) Nanoparticle characterization by cyclical electrical field-flow fractionation. Anal Chem 83(17):6565
24. Latham AH, Freitas RS, Schiffer P et al (2005) Capillary magnetic field flow fractionation and analysis of magnetic nanoparticles. Anal Chem 77(15):5055–5062
25. Beveridge J S. (2012) Differential magnetic catch and release: separation, purification, and characterization of magnetic nanoparticles and particle assemblies. Dissertations and theses-Gradworks
26. Beveridge JS, Buck MR, Bondi JF et al (2011) Purification and magnetic interrogation of hybrid Au–Fe_3O_4 and FePt–Fe_3O_4 nanoparticles. Angew Chem Int Edit 50(42):9875–9879
27. Yavuz CT, Mayo JT, William WY et al (2006) Low-field magnetic separation of monodisperse Fe_3O_4 nanocrystals. Science 314(5801):964
28. O'connell MJ, Bachilo SM, Huffman CB et al (2002) Band gap fluorescence from individual single-walled carbon nanotubes. Science 297(5581):593
29. Guo Z, Fan X, Xu L et al (2011) Shape separation of colloidal gold nanoparticles through salt-triggered selective precipitation. Chem Commun 47(14):4180

30. Sharma V, Park K, Srinivasarao M (2009) Shape separation of gold nanorods using centrifugation. PNAS 106(13):4981–4985
31. Jamison JA, Krueger KM, Yavuz CT et al (2008) Size-dependent sedimentation properties of nanocrystals. ACS Nano 2(2):311–319
32. Svedberg T, Nichols JB (2002) Determination of size and distribution of size of particle by centrifugal methods. J Am Chem Soc 12:2910–2917
33. Li P, Huang J, Luo L et al (2016) Universal parameter optimization of density gradient ultracentrifugation using CdSe nanoparticles as tracing agents. Anal Chem 88(17):8495

Chapter 2
Basic Concepts of Density Gradient Ultracentrifugation

Jindi Wang, Jun Ma and Xuemei Wen

Abstract This chapter reviews the development of centrifugal equipment and its achievements in various fields, especially in the field of life sciences. Based on the development of the equipment, centrifugal technology also plays a very crucial role in scientific research. Afterward, we introduce three common types of separation techniques: differential centrifugation, rate-zonal centrifugation, and isopycnic separation. For each type, we describe the basic principles, characteristics, and scope of application in detail.

Keywords Density gradient ultracentrifugation · Revolution of centrifugation Differential separation · Isopycnic separation · Rate-zonal separation

2.1 Revolution of Density Gradient Ultracentrifugation

Centrifugal technology has been developed for centuries. As early as the ancient times, people utilized centrifugal force by using tied rope to circularly move pot to squeeze honey or slurry. The dehydration and separation processes from textile and light industry in mid-eighteenth century facilitated the appearance of modern model of the centrifuge. In 1836, the first three-foot centrifuge came out in Germany, and a cotton dehydrator also appeared. In 1878, the Swedish de Laval developed a manual-belt-type milk separator and used it to make cream. Six years later, he invented the butter separator driven by the diesel engine to further improve the separation efficiency. These types of equipment greatly promoted the development of the cream industry. During this period, with the rapid development of industry, centrifugal technology has also made great progress. However, due to the limitations of science and technology at that time, high-speed centrifugal equipment could not be realized, which greatly hindered the application of separation technology in the field of basic science.

Since the twentieth century, the industrial manufacturing techniques have been greatly improved. High-speed centrifuges and ultrahigh-speed centrifuges were produced one after another. In the 1920s, US company "DuPont" produced the oil turbine centrifuge. In 1933, the air turbo centrifuge was developed by using the

X. Sun et al., *Nanoseparation Using Density Gradient Ultracentrifugation*,
SpringerBriefs in Molecular Science, https://doi.org/10.1007/978-981-10-5190-6_2

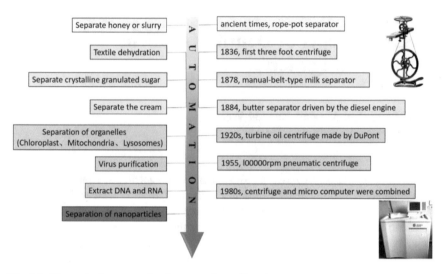

Fig. 2.1 Shows the important development of centrifugation

compressed air to drive the worm wheel and then drove the centrifuge to rotate. In 1955, US company "Beckman" launched up to 100,000 rpm pneumatic centrifuge; in the late 1970s, variable frequency motors emerged. Then, in the 1980s, micro-computers were integrated with the variable frequency motors, so the speed and performance of the centrifuge were highly improved. The continuous improvement in centrifuge extends and strengthens the exploration in many scientific fields, especially in the life sciences. As shown in Fig. 2.1, we have summarized some representative developments of centrifugation.

As the research of cells gradually goes in-depth, the study of subcellular structures has become a hot topic in life sciences, which demands the pure orga-nelles from cell. Hence, scientists began to use the centrifugal method to obtain organelles from the cell brei. The first breakthrough was made in 1934, and Bensley and Hoerr [1] used high-speed centrifugation to isolate mitochondria from the cell homogenate. Consequently, they studied the chemical composition and physio-logical function of mitochondria and concluded that mitochondria are the center of cell oxidation. In that case, biology had ushered in a new development period.

Extending from prior successes, Hogeboom et al. [2] put forward a complete separation protocol describing how to separate various suborgans from the cell homogenate in detail. This protocol is also represented as the preliminary formation of differential centrifugation model.

The differential centrifugation method, also known as pelleting separation, which is a method of using a certain speed centrifugation to separate materials from homogeneous suspension, has been widely applied to separate and purify nanoparticles by centrifuge redispose circles. In the 1950s, Brakke set up a rate-zonal separations method. He proposed the concept of density gradient centrifugation for the first time [3] and demonstrated this new method for virus purification. On the basis of this method, Meselson et al. [4] proposed a balanced gradient centrifugation method, with the gradient distribution of DNA in CsCl. De Duve's group [5] used density gradient centrifugation method for the first time to get the lysosome. This work represented another major leap in human under-standing of the cell; Duve also won the 1974 Nobel Prize in physiology or medi-cine. This series of meaningful works made the density gradient centrifugation attract more and more scientific attentions. After that, the centrifuge technology had been greatly improved.

Now, density gradient centrifugation technique is widely used to study the biological characteristics and the separation of biological samples on cells, sub-cellular organelles, nucleic acids, proteins, enzymes, and receptors. It has become one of indispensable technical means in life sciences. With the rapid development, density gradient centrifugation technique not only plays an increasingly great role in biological field, but also promotes the continuous technology improvement and development in broader application prospects.

2.2 Differential Centrifugation

2.2.1 Introduction

Differential centrifugation is a centrifugal method based on the different settling velocity of nanoparticles in a homogeneous medium. As shown in Fig. 2.2, the technique is accomplished by repeatedly centrifuging a suspension from low speed to high speed to achieve separation. In a typical process, different sized particles move toward the bottom of the centrifuge tube with different settling rates and eventually precipitate. Specifically, in order to separate the different components, we can adjust the rotation speed to change the applied centrifugal force and further to manipulate the settling velocity of the particles. This centrifugation is also called differential pelleting, because the samples can be separated by sequentially increasing the centrifugal speed to obtain a series of precipitations and supernatants.

Next, we give a brief description of how to select the centrifugal speed. As shown in Fig. 2.2, firstly, low speed can be used to make the largest particles settle at the bottom of the centrifuge tube, which are precipitation 1. After removing the precipitate 1 and resuspension, a medium speed should be used to obtain the medium particles. Finally, the precipitation of the smallest sized particles needs the highest centrifugation speed. Through the above steps, the preliminary separation of different components can be realized. In particular, it can be seen from the tubes that the fractions obtained by this approach are not strictly monodispersed.

The applications of differential centrifugation are widespread, especially in biochemistry for the separation, extraction, and enrichment of bioactive substances, such as animal/plant virus and subcellular components. Figure 2.3 shows a typical

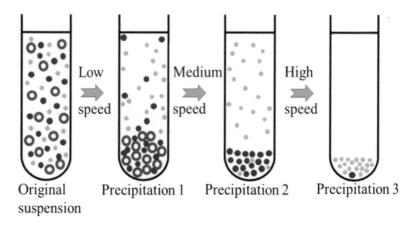

Fig. 2.2 Basic process of differential centrifugation

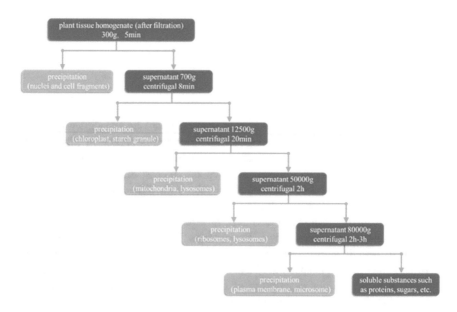

Fig. 2.3 Differential separation process of subcellular structures of plant materials [3]

differential separation process of subcellular structures of plant materials. Samples got from the precipitation must go through the resuspension, recentrifugal, and rewashing processes to get much purer samples of the particles. The yield by differential centrifugation is not high. Except for the slowest settled particles in the group, however, it is hard to get a pure component.

2.2.2 Basic Principle

When the particles are settled in centrifugal field, their sedimentation speed depends on many parameters, such as mass, size, shape of particles, and density, viscosity of media. The detailed information will be discussed in Chap. 4. Here, we will give basic principles of differential separation.

For a particle (suspended in a liquid of known density and viscosity) of a given size and density, the density and viscosity of the liquid medium are known quantitatively, and for a certain particle, r, ρ_p, η, and f are also known quantitatively, and the sedimentation coefficient (s) can be calculated as $s = \frac{2r^2(\rho_p - \rho_m)}{9\eta(f/f_0)}$ and

$$\frac{dx}{dt} = s\omega^2 x$$

For convenience, we mark the sedimentation coefficient as follows

$$s = \frac{dx/dt}{\omega^2 x}$$

Accordingly, the sedimentation coefficient is the ratio of particle velocity (dx/dt) to its acceleration ($\omega^2 x$), which determines the sedimentation rate of particles, and is of great importance for the sedimentation behavior of particles in differential separation.

2.2.3 Applicable Conditions

In a typical process of differential separation, all the different sized particles move toward the bottom of the centrifuge tube. As mentioned above, the differential separation is based on the different settling velocity of nanoparticles in a homogeneous medium and determined by the sedimentation coefficient.

For an intuitionistic comparison, three types of particles with 100S, 10S, and 1S ($S = 10^{-13}s$, s is sedimentation coefficient) are mixed and sorted by differential separation, and the separation results are quantitatively analyzed, as shown in Fig. 2.4.

It is obvious that, after the optimization of differential separation parameters, there are still 1% of smallest sized particles and 10% medium-sized particles precipitated with largest particles. When applying the precipitation process again, there are still some big-sized precipitates mixed in small-sized precipitates [7]. Washing precipitation can improve the purity of separation, and this process can resuspend and centrifuge the precipitation to improve the efficiency of differential

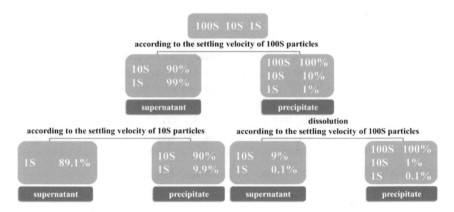

Fig. 2.4 Separation result of the particles of three sedimentation coefficients in the sample: 100S, 10S, and 1S

centrifugation. By using this method, the maximum-sized precipitation with high purity can be obtained after several washing.

Thus, the differential separation method is suitable for the samples which contain big weight difference between the fragments, or the difference between sedimentation coefficients (the measurement on settlement, which is defined as the quotient of a particle's sedimentation velocity over the acceleration) of mixed samples should be more than ten times.

2.2.4 Basic Calculation

Angular rotor is often used in differential centrifugation method, and the particles are deposited on the outside wall of centrifuge tube as moving toward the bottom (Fig. 2.5) [8].

According to the definition of sedimentation coefficient:

$$s = \frac{dr/dt}{w^2 r}$$

we can get:

$$s \int_{t_1}^{t_2} w^2 dt = \ln \frac{r_2}{r_1}$$

Centrifugal time is the centrifuge to run from start to run completely stop, and the expression is:

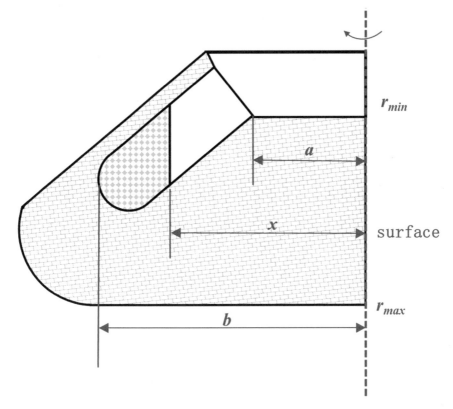

r_{min}

a

x

surface

r_{max}

b

Fig. 2.5 Movement of the interface in the angular rotor. *a* Minimum centrifugal radius; *b* maximum centrifugal radius; *x* the distance of the particle mobile interface

$$\int_{t_0}^{t_{stop}} w^2 dt = \int_{t_0}^{t_{max}} w^2 dt + \int_{t_{max}}^{t_{off}} w^2 dt + \int_{t_{off}}^{t_{stop}} w^2 dt$$

The time of the acceleration and deceleration is different among kinds of centrifuge. The actual effective expression is as follows:

$$t = \frac{1}{3}(t_{max} - t_0) + (t_{off} - t_{max}) + \frac{1}{3}(t_{stop} - t_{off})$$

Or

$$t = \frac{1}{3}t_+ + t_{constant} + \frac{1}{3}t_-$$

If *s* value of the particle is known, the whole matter of the size will be settled down (Fig. 2.6).

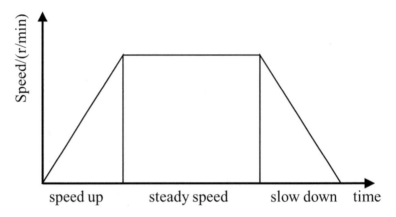

Fig. 2.6 Scheme of actual effective centrifugal time calculation

$$b = ae^{sw^2t}$$

$$\ln\frac{b}{a} = sw^2t$$

Known

$$w = 2\pi n$$

For any r and t, the following two equations are used:

$$n^2t = \frac{3600}{4\pi^2s}\ln\frac{b}{a} \times \frac{1}{n^2}\,(s)$$

$$n = \sqrt{\frac{3600}{4\pi^2s}\ln\frac{b}{a} \times \frac{1}{t}}$$

2.3 Rate-Zonal Separation

2.3.1 Introduction

Rate-zonal separation is a kind of centrifugation method using zone rotor [9], which is based on the differences in sedimentation velocities of samples in density gradient [10]. The method is to put the samples on the top of the density gradient which has a smaller scale of density ($\rho_{media} < \rho_{particle}$) and mildly changed slope. Mass and the viscous resistance are different for the particles with different shapes and sizes under a given centrifugal force, which leads to different sedimentation coefficients and consequent different particle sedimentation velocities, and effective separation.

Typical zone rotor is a hollow pot body, including upper and lower parts, which is convenient to clean and maintain. It cuts down the amount of centrifuge tubes and greatly increases the use of volume. The center of rotor is axis and partition, and the rotor cavity is divided into several regions by partition, which can reduce the eddy current and stabilize gradient liquid. Zone rotors are divided into redirect, non-directional, continuous flow, etc. Redirect rotor is that gradient medium and samples in the rotor present radial in the process of centrifugal acceleration, and slow down the process of orientation change according to the gravity, gather gradient zones from the top or the bottom after centrifugal separation. Continuous flow zone rotor is that samples or medium are added through the rotation of coaxial rotor rotating sealing device when the machine is running. After the completion of the first centrifugation, centrifugation is further running through the same rotating sealing device of continuous pumping medium to replace the separation zone. Also, the isolated samples are collected through monitor testing part.

Usually, by using zonal rotors and swing-out rotors, and taking advantage of their long settlement path, the particles can be fully separated. Sucrose and iodinated gradient media are mostly used. The centrifugation process should be stopped when the fastest settling particles reach the bottom of the rotor or wall of tube. Rate-zonal separation undertakes in strong centrifugal field and costs short period of separation time (generally not more than 4 h). The sample particles would go through the gradient and stack at the bottom of the rotor if the centrifugal time is too long. In order to avoid such situation, a cushion layer whose density is greater than the maximum grain density is required to be put at the bottom of the gradient.

In homogeneous medium, the interactions between the particles and medium should be the same in the process of settlement. Under a given centrifugal field, the settling velocity of particles increases with the increase of the centrifugal radius. In that case, the centrifugal radius of leading particles would increase faster than that of slow ones, and consequently, settling velocity of leading particles would also increase faster, leading to the enhanced resolution. While in the density gradient, density and viscosity of medium continuously increase, and leading particles would enter into the higher density area first. Thus, sedimentation velocity of leading particles will slow down quickly due to the strong resistance. The greater the slope of gradient, the greater the speed goes down, which will be harmful to improve separation resolution. However, if there is no density gradient, all particles would be settled at the end of centrifuge tubes. At the same time, the convection produced by the change of the rotor speed can mix separated particles again in homogeneous medium [11]. Hence, in order to obtain maximum separation resolution, the samples suspension added on the gradient should be as thin as possible, viscous resistance and buoyancy of particles in a given centrifugal field by gradient moving should be minimal, and the separation should be carried out as soon as possible after the formation of density gradient to minimize the diffusion effect (Fig. 2.7).

To guarantee the stability of sample zone, the sample weight added in the density gradient liquid column should be matched with the concentration of the

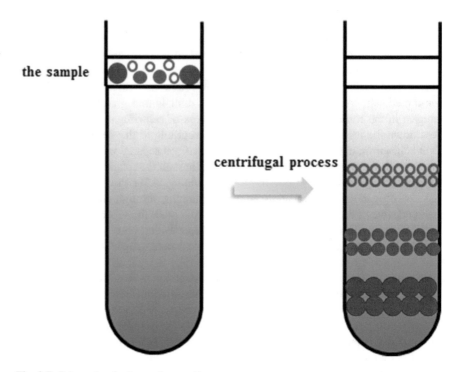

Fig. 2.7 Schematic of rate-zonal separation

sample. The maximum density of gradient (usually the bottom gradient) must be lower than that of particles, and the minimum density of gradient (usually the top gradient) must be greater than the density of sample suspension. If the subsidence coefficient of different particles is similar, gradient density with smaller range should be chosen, and if the differences of particle density are large, the steep gradient will be preferred [12]. If the sample concentration is too high, the zone will be severely spread; while if the sample concentration is extremely low, it would be hard to identify the sample zone [13].

2.3.2   Basic Principle

The particle's settling velocity in the centrifugal field is influenced by the density difference between the particle and surrounding medium, particle size (e.g., diameter and length), centrifugal field force, and viscous drag (associated with the viscosity and surface area of the particle). The calculation can be expressed by the following equation

$$v = kgr^2 \frac{d_1 - d_2}{\mu}$$

v Particle settling rate (cm/s).
g Gravity acceleration.
r The radius of the particle (cm).
d_1 The specific gravity of particles.
d_2 Specific gravity of water medium.
μ Viscosity of water medium.
k The shape factor varies with shape

The speed of any particle (dr/dt) goes up with the increase of the rotor radial distance r (i.e., increases with r). It is obvious that the dr/dt in the density gradient is not depended on r, and it changes with the density and viscosity of the medium. In the later discussion of the method, the changing in the settlement velocity with the density of the medium will be unveiled, and the estimated settling time under different conditions will be demonstrated. The basic principle of particle subsidence is the same as the differential velocity centrifugation.

2.3.3 Considerations

All in all, there are some guidelines that need to be followed in rate-zonal centrifugation. Firstly, the density of particle suspension must be lower than that of the top gradient. Then, the density of the sample particles must be higher than that of bottom gradient. In addition, the path length of the gradient must be long enough to sufficiently allow the particles to be separated. Finally, the centrifuge should be stopped before or as long as the leading particles reach the bottom of the gradient.

2.3.4 Applicable Conditions: High Density Nanostructures

Rate-zonal separation is a dynamic method related to particle size and density. Buoyancy and the viscous resistance are different for the particles with various shapes and the size in the density gradient, which leads to different sedimentation rates, so under the condition of a certain centrifugal, particle size, morphology, and structure can be effectively separated. Figure 2.6 shows process of rate-zonal separation. Compared with isopycnic separation, rate-zonal separation can be used for separating these particles, density of which is much higher than the gradient medium of colloidal particles, such as metal nanoparticles, which greatly expands the scope of separation. And the time of rate-zonal separation (such as 15 min) is shorter than typical isopycnic separation time (usually 12 h). In addition, parameters for the separation, such as time, centrifugal velocity, and gradient density, can be adjusted specifically for the separated particles.

The rate-zonal separation method has been successfully applied to sort the various colloidal nanostructures, such as FeCo@C, Au nanoparticles, graphene, and CdS nanorods [14].

2.4 Isopycnic Separation

2.4.1 Introduction

When the density of particles suspended in the mixture suspension is different from that of the gradient medium, under the effect of the centrifugal force, they either settle or rise along with the gradient. The particles will keep moving until they reach the same density of gradient. At this point, the particles are in the minimized energy state [15]. Here, the gradient medium density is equal to the buoyancy density of particles, which is considered as "isopycnic state." This centrifugal method, which establishes and measures the equilibrium state, is considered as isopycnic equilibrium gradient centrifugation. The effect of the density isopycnic centrifugal separation depends on the density difference between the particles and separation ability of gradient medium. The particles with bigger density difference are easier to separate since the separation effect is regardless of the size and shape of the particles for isopycnic one. However, the size and shape have a profound effect on the equilibrium time and the width of the zone. In isopycnic equilibrium gradient centrifugation, the buoyancy density of particle is not constant and highly related to the density of the particle and hydration degree, and it also relates to these factors such as gradient media and interactions. For example, some particles are easy to be hydrated, so their net density will be lowered; sometimes the net density would be even changed due to the replacement of surface structural water by small molecules of gradient medium. The influences of these factors should be taken into consideration when colloidal nanoparticles are separated.

In practical, sample should be loaded on preformed gradient with increasing density before centrifuge starting; after long term centrifugation (usually longer than 12 h), the particles would find the layer with same density to stop and locate therein (Fig. 2.8).

The conditions of isopycnic equilibrium gradient centrifugation depend on the requirements, properties, medium, gradient characteristics of experiment [16].

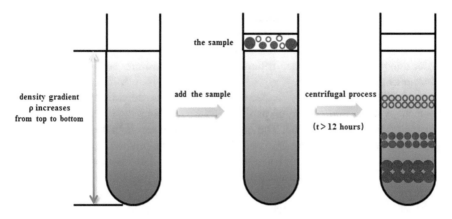

Fig. 2.8 Schematic of isopycnic separation

2.4.2 Basic Principle

For the physical balance in centrifugal field, concentration distribution can be calculated according to the following formula:

$$\frac{dc_m}{dr} \times \frac{1}{c_m} = \frac{\omega^2 r_\theta M(1 - \bar{v}\rho s)}{RT}$$

Assuming that the relationship of the density and gradient concentration distribution is:

$$\rho_s(r) = \frac{1}{\bar{v}} + (r - r_\theta) \times \frac{d\rho_s}{dr}$$

when density balance particle concentration distribution relationship is:

$$\ln\frac{c_\theta}{c_r} = \frac{\omega^2 r_\theta M\bar{v}(d\rho_s/dr)(r - r_\theta)}{2RT}$$

r_θ Radius of zonal density, corresponding to the particle equilibrium average density

c_r The concentration of the radius

A linear relationship between $\ln c_r$ and $(r - r_\theta)^2$, intercept is $\ln c_\theta$, slope is $\omega^2 r_\theta M\bar{v}(d\rho s/dr)/(2RT)$.

Conditions of isopycnic equilibrium gradient centrifugation are confirmed according to the requirement of the experiment, sample characteristics, the characteristics of the medium and gradient.

Aqueous soluble gradients are used in isopycnic equilibrium gradient centrifugation, and pH value depends on the nature of the biological sample. In most of the cases, sucrose, glucose, glycerol, sorbitol, tartaric acid, potassium bromide, cesium chloride, potassium chloride, trifluoroacetic acid, cesium, and dextran can be chosen as gradient medium.

According to the density difference between particles, the particles will not move to form a layer upon layer of particles when reaching their respective density regions in gradient separation. The equal density gradient includes the preforming gradient and the self-forming gradient. The self-forming gradient is that the gradient medium mixed with the samples together, sample particle distribution in different density position after centrifugation. The preforming gradient is a way that the gradient liquid is placed in a centrifuge tube according to the differences between light and weight. It includes linear gradient and step gradient. The former is prepared by the gradient instrument or by step gradient, which is formed by diffusion, and the latter is formed by manual placement.

If the maximum density of the gradient medium is higher than the maximum density of the sample particles, or the sample density is between the maximum and the minimum of the gradient density, the sample cannot reach to the bottom of the centrifuge tube. The density gradient range includes the density required to separate the particles. When the particles are separated in the process of centrifugation and settle to a position with equal density, the zone is integrated. In this integrated zone, the density of the gradient liquid is as same as the density of the sample. The effect of such density gradient centrifugation is determined by the density differences of the sample particles. The greater the difference in sample density is, the better the separation effect would achieve [17].

2.4.3 Characteristics

The characteristics of isopycnic separation:

- Its resolution is high and can separate two particles whose density differences are less than $0.01 \sim 0.02$ mg/cm^3 ($1/10^5$ density of water).
- It can resist the disturbance caused by temperature change and acceleration and deceleration. Its anti-convection and anti-disturbance ability is excellent.
- It can handle a large amount of sample simultaneously.
- The sample is suspended in a gradient medium and does not form precipitation, which is beneficial for maintaining biological activity.
- The resolution is related to the profile of terminal density gradient after centrifugation, which is mainly affected by the medium types and centrifuge speed. Extending the time of acceleration and deceleration would help to preserve the separation accuracy.

2.4.4 Applicable Conditions: Low-Density Nanostructures

Isopycnic separation is a method of using density gradient or over-speed centrifugal which depends on the density of the separated material. From the separation of biological macromolecular, to the separation of carbon nanotubes, isopycnic separation has achieved great success. Particles move to the positions where the medium density matches with the density of each colloidal particle under given centrifugal field to separate. There is no relationship between separation effect and the size of the particles; however, separation effect mainly depends on the density differences between particles and gradient medium [18].

Density gradient centrifugation appears to be the most effective in separating nanotubes of smaller diameters, such as carbon nanotubes. Mark C. Hersam's group reported that single-walled carbon nanotubes can be separated in the density range of $1.11–1.17$ g cm^3 [19]. Graphene nanosheets with surfactant wrapping are separated, and their net densities are 1.1 g/cm^3 [20]. Isopycnic density gradient centrifugation method reaches a limitation when it is extended to the separation of metal nanoparticles. Such a method requires that the components for separation have densities within a gradient range. Aqueous density gradient media usually have densities less than 1.4 g/cm^3, which is much lower than the density of metal nanoparticles. Therefore, this method is only suitable for the separation of colloidal particles with low density [21]. In order to expand the scope of separating particles, other density gradient centrifugation methods are required.

References

1. Bensley RR, Hoerr NL (1934) Studies on cell structure by the freezing-drying method VI. The preparation and properties of mitochondria. Anat Rec 60(4):449–455
2. Hogeboom GH, Schneider WC, Palade GE (1948) Cytochemical studies of mammalian tissues; isolation of intact mitochondria from rat liver; some biochemical properties of mitochondria and submicroscopic particulate material. J Biol Chem 172(2):619
3. Brakke MK (1951) Density gradient centrifugation: a new separation technique. J Am Chem Soc 73(4)
4. Meselson M, Stahl FW, Vinograd J (1957) Equilibrium sedimentation of macromolecules in density gradients. Proc Natl Acad Sci USA 43(7):581–588
5. De Duve C, Baudhuin P (1966) Peroxisomes (micro bodies and related particles). Physiol Rev 46(2):323
6. Jacobus WE, Vandegaer KM, Moreadith RW (1986) Aspects of heart respiratory control by the mitochondrial isozyme of creatine kinase. In: Myocardial and skeletal muscle bioenergetics. Springer, USA
7. Arnold MS, Stupp SI, Hersam MC (2005) Enrichment of single-walled carbon nanotubes by diameter in density gradients. Nano Lett 5(4).713–718
8. Arnold MS, Green AA, Hulvat JF, Stupp SI, Hersam MC (2006) Sorting carbon nanotubes by electronic structure using density differentiation. Nat Nanotechnol 1(1):60
9. Green AA, Hersam MC (2009) Processing and properties of highly enriched double-wall carbon nanotubes. Nat Nanotechnol 4(1):64–70

10. Khripin CY, Tu X, Heddleston JM, Silvera-Batista C, Hight Walker AR, Fagan J et al (2013) High-resolution length fractionation of surfactant-dispersed carbon nanotubes. Anal Chem 85 (3):1382–1388
11. Schlegel RHJ (2008) Differential centrifugation. Encyclopedia of genetics, genomics, proteomics and informatics. Springer, Netherlands
12. Rickwood D (ed) (1984) Centrifugation: a practical approach, 2nd edn. IRL Press
13. Dorn AR (1990) Cell separation process. US patent US4927750
14. Schneider WC, Hogeboom GH (1951) Cytochemical studies of mammalian tissues: the isolation of cell components by differential centrifugation: a review. Can Res 11(1):1–22
15. Hayakawa M, Otoguro M, Takeuchi T, Yamazaki T, Iimura Y (2000) Application of a method incorporating differential centrifugation for selective isolation of motile actinomycetes in soil and plant litter. Antonie Van Leeuwenhoek 78(2):171–185
16. Frühling J (1979) Exchange of cholesterol between the subcellular fractions during differential centrifugation of rat adrenocortical tissue. In: Separation of cells and subcellular elements, pp 91–93
17. Patsch JR, Sailer S, Kostner G, Sandhofer F, Holasek A, Braunsteiner H (1974) Separation of the main lipoprotein density classes from human plasma by rate-zonal ultracentrifugation. J Lipid Res 15(4):356
18. Groot PH, Scheek LM, Havekes L, van Noort WL, van't Hooft FM (1982) A one-step separation of human serum high density lipoproteins 2 and 3 by rate-zonal density gradient ultracentrifugation in a swinging bucket rotor. J Lipid Res 23(9):1342–1353
19. Arnold MS, Stupp SI, Hersam MC (2005) Enrichment of single-walled carbon nanotubes by diameter in density gradients. Nano Lett 5(4):713–718
20. Kuang Y, Song S, Huang J, Sun X (2015) Separation of colloidal two dimensional materials by density gradient ultracentrifugation. J Solid State Chem 224:120–126
21. Arnold MS, Green AA, Hulvat JF, Stupp SI, Hersam MC (2006) Sorting carbon nanotubes by electronic structure using density differentiation. Nat Nanotechnol 1(1):60–65

Chapter 3
Density Gradient Ultracentrifugation Technique

Qian Zhang and Xiong Sun

Abstract As a general, non-destructive, and scalable separation method, DGUC has recently been demonstrated as an efficient way of sorting colloidal nanoparticles according to their differences in chemical, structural, size, or morphology. After the introduction of basic concepts of density gradient ultracentrifugation, for the practical applications, there are various parameters to be considered. Nanoparticles will have different movement ways in different separation systems. In principle, particle movement characteristic in liquid media not only depends on the centrifugal force but also relies on the density, size, and shape of particle and the density and viscosity of the liquid medium and so on, while the gravity and intermolecular force can be ignored. In this chapter, typical parameters such as choice of gradient media, density gradient, rotor type, centrifugation speed, and time will be discussed.

Keywords Gradient media · Step gradient · Continuous gradient
Rotor type · Centrifugal force · $\omega^2 t$ calculation method
K' coefficient estimation method

3.1 Choice of Gradient Media

Up to now, only a few types of liquids or solutions are utilized to form density gradients to separate biomolecules or nanoparticles (NPs). Since NPs are not biologically active, harsher conditions beyond aqueous solutions could also be used to separate NPs.

3.1.1 Gradient Media

Media applied in density gradient centrifugation includes small hydrophilic molecular organic compound (ethanol, ethylene glycol, glycerin, etc.), macromolecule organic compound (polysaccharide, protein, etc.), alkali metal salts (Cs, K, Rb, Na

© The Author(s) 2018
X. Sun et al., *Nanoseparation Using Density Gradient Ultracentrifugation*,
SpringerBriefs in Molecular Science, https://doi.org/10.1007/978-981-10-5190-6_3

salts, etc.), colloidal silica (Percoll, Ludox, etc.), and three organic toluene derivatives (covering Metrizamide, Nycodenz, Iodixanol, etc.) [1].

Clearly, solution prepared from aforementioned media differs in density, viscosity, concentration, chemical composition, and properties. As a result, solution made from one to two kinds of media chosen above can be used to separate colloidal NPs. Before separation, several important points should be taken into consideration when choosing the ideal gradient media:

① For the nanoparticles to be separated, the gradient media should be inert to colloidal samples under ambient condition. Moreover, it had better has no toxicity.

② The gradient media ought to keep the dispersibility of the separation system without causing agglomeration in the suspension.

③ The physical and chemical properties of gradient media should be known, such as the exact density, viscosity, and the specific mixed ratio of the gradient media.

④ The gradient media should be pH neutral, and its ionic strength should be low enough to avoid possible aggregations of the colloidal NPs.

⑤ The gradient media should be colorless which absorb out of the UV range of the spectrum (i.e., <400 nm).

⑥ The gradient media should be easily removed from the separated particles without loss of the activity of the sample.

⑦ The centrifugal media should possess bactericidal action, remain its separation characteristics after separating, and do not corrode centrifugal tube, the rotor, and other relevant equipment.

⑧ The gradient media should be pure substance with low price.

The above are the requirements of the ideal gradient media, while no single liquid can meet all these criteria. In practical work, the selection of proper gradient media should be considered based on the target structure of NPs [2].

3.1.2 Commonly Applied Gradient Media

3.1.2.1 Aqueous Gradient Media

Water is by far the most widely investigated polar solvent and is described as the universal solvent for its capability to dissolve majorities of substance, environmentally benignity, earth-abundant storage, and low cost. Factually, quite a great number of hydrophilic media constitute the aqueous gradient system with the maximum density range of 1.4 g/cm^3. In the pursuit of superior separation effect of various inorganic nanoparticles in aqueous circumstances, sucrose, glycerol, and iodinated compounds are extensively utilized.

Table 3.1 Effect of temperature on the viscosity of sucrose solution

Concentration (mass concentration)/%	Viscosity/mPa·s				
	0 °C	5 °C	10 °C	15 °C	20 °C
20	3.77	3.14	2.64	2.55	1.95
30	6.67	5.42	4.48	3.76	3.19
40	14.58	11.45	9.16	7.46	6.16
50	44.74	33.16	25.17	19.52	15.42

Sucrose

Sucrose solution is the most universally used aqueous gradient medium by far, along with its cost-effective virtue. Sucrose is easily soluble in water, and the density range can be tuned from 1.00 to 1.32 g/cm^3 and could be further enlarged by adding sodium or potassium bromide, citrate, or tartrate into the solution. Generally, it is easy to prepare sucrose solutions in a variety of buffer systems, and the concentration of sucrose solutions has a definite relationship with its viscosity, density, and refractive index. With the assistance of sucrose gradient, tremendous colloidal nanomaterials have been separated by diameter, including magnetic NPs (cobalt ferrite and manganese ferrite), $CuInS_2$ quantum dots, graphene [3–5].

Furthermore, the viscosity of sucrose solutions is highly dependent on the temperature and concentration (as shown in Table 3.1) [6]. Increasing viscosity of the sucrose solutions can be reduced by replacing H_2O with D_2O which has a density of 1.11 g/cm^3.

Commercially available sucrose generally contains several impurities, so sucrose should be purified first to ensure that it is suitable for density gradient work. Moreover, specially refined sucrose with high purity is broadly available.

Glycerol

Glycerol is miscible with water and easy to volatilize, so that it can be directly removed after the separation. Glycerol is half of sucrose in density and is convenient to prepare solutions with high concentration. Usually, 10–30% linear glycerol gradient can be used as an alternative to the 5–20% linear sucrose gradient for the separation.

However, physio-chemical characteristics of glycerol are less well known compared with sucrose, and the high viscosity of glycerol hinders it from wide use for density gradient.

Iodinated Compounds

Iodinated organic compounds, which possess merits such as low toxicity, low osmolality, high solubility, and high density, were firstly employed as density gradient in 1953 by Holter's group. Compared to other gradient media, it is generally believed that the triiodobenzoic derivatives should be the most ideal and widely adapted gradient media up to now.

Non-ionic fraction of the triiodobenzene derivatives containing Metrizamide, Nycodenz, and iodixanol is suitable for the density gradient separation of all particles due to excellent compatibility and stability for NPs. Compared with the above, iodixanol ($C_{35}H_{44}I_6N_6O_5$) has a relatively larger formula weight and lower viscosity; thus, its density gradient is readily formed and most widely used in separating carbon-based materials [7–13]. Isosmotic solution with diverse concentrations can be prepared using 60% w/v iodixanol aqueous solution (purchased as Optiprep^{TM}, 1.32 g/cm^3).

Ethanol

Ethanol (EtOH), which is also called alcohol or ethyl alcohol, is the principal type of alcohol found in alcoholic beverages. As well known, it is a volatile, flammable, colorless liquid with a slight characteristic odor.

Solutions of ethanol in water with different volume ratios (e.g., 20, 30, 40, and 50%) can be used as the density gradient, which, more importantly, can even provide the hydrophilicity difference to capture NPs with different hydrophilicity/polarity. For instance, Deng et al. [14] developed a hydrophilicity gradient ultra-centrifuge separation technique based on solubility/stability difference of carbon nanodots (CDs) in gradient of ethanol/water solutions to separate CDs.

Ethylene Glycol

Regarded as the simplest diol, ethylene glycol (EG) is moderately toxic and miscible with water. The density gradient based on different concentrations of EG/water solutions has been widely applied to separate NPs, such as Ag nanoplates (NPTs) [15], Pd nanoplates, and the Au nanoparticles [16]. Besides, Chang et al. [17] successfully sorted the colloidal MgAl LDHs nanosheets and NiAl LDHs nanosheets by their lateral size using the EG aqueous gradient, indicating the potential of extending the density gradient ultracentrifugation separation (DGUS) method for the further study of structure, composition, and properties of multi-component nanomaterials.

Inappropriate Media for Nanoseparation:

Alkali Metal Salts

In addition to the above mentioned, gradients using alkali metal salts, including sodium bromide (NaBr), sodium iodide (NaI), cesium chloride (CsCl), cesium bromide (CsBr), are widely used for the separation of biological materials (e.g., DNA, RNA, nucleic acid), while for the separation of inorganic NPs there are only a few cases using the CsCl solution [18–20] which can reach 1.92 g/cm^3 in density as a gradient media, since salts in density gradient media will reduce the electrostatic stability of charged NPs in liquid solutions and induce the aggregation [16].

3.1.2.2 Organic Gradient Media

The separation of biomolecules usually needs strict osmotic and viscous liquid environment, namely aqueous gradient. While a lot of nanoparticles were synthesized in the organic phase and got solubilized therein (such as Au, CdSe, and Si nanocrystals), phase transfer might cause serious aggregation. In addition to extending the range of colloidal systems which can be separated, there are several other advantages of using an organic density gradient rather than the conventional aqueous gradients. To begin with, combining synthesis optimization and separation together can prepare samples that cannot be prepared by synthesis optimization alone. Secondly, colloidal NPs synthesized and dispersed in an organic medium can be directly separated after synthesis without transferring to an aqueous medium, which avoids the possible aggregation and clustering of NPs under non-optimized conditions. Thirdly, since the density gradients are composed of organic solvents without any solid additives, the solvents can be evaporated without leaving any residue, to get "pure" sample. Colloidal NPs with constant size distribution can thus be captured in the gradient together with polymers.

Cyclohexane/Carbon Tetrachloride

Cyclohexane is a cycloalkane with the molecular formula of C_6H_{12} and a normal density of 0.779 g/cm^3, which is miscible with various organic solvents and frequently used as a non-polar organic gradient media in density centrifugation. Carbon tetrachloride (CCl_4) is a colorless and volatile liquid with a slight sweet smell and a normal density of 1.595 g/cm^3. The density gradient based on the combination of non-polar cyclohexane and CCl_4 can realize a large density range, which is ideal for the separation of NPs [21–23].

Chlorobenzene/2,4,6-Tribromotoluene

Chlorobenzene is a colorless liquid with a density of 1.484 g/cm^2, and the mixture of chlorobenzene and 2,4,6-tribromotoluene (TBT) is usually used as a density gradient to separate carbon nanotubes [24] and monodisperse silicon nanocrystals [25].

Other macromolecular organics such as Ficoll and dextran are also suitable for fractionating biological materials, while the difficult handing procedure hinders the applications due to their relative low density and high viscosity.

3.2 Choice of Density Gradient

In order to obtain the maximum resolution, the suspension samples on gradient should be as thin as possible. In a given centrifugal field, when particles move through the density gradient, the buoyancy density and viscous resistance should be minimal, and the gradient density range should be small [26]. In this case, the centrifugal separation time will be shorter and there will be minimum diffusion effect. Gradients can be divided into continuous and discontinuous (step) gradients, as illustrated in Fig. 3.1.

3.2.1 Step Gradient

Discontinuous (step) density gradient is primarily restricted to isopycnic separations where it obtains sharper bands.

The preparation method for step gradients is quite simple. A series of solutions with increasing density are prepared, and orderly add these solutions to the centrifuge tube or zonal rotor by using the liquid transfer tube, syringe, or density

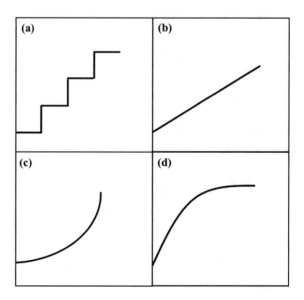

Fig. 3.1 Models of different types of gradients. Step gradients **a** are used only for isopycnic separations. Continuous gradients are **b** linear, **c** convex, **d** concave and could be used for rate-zonal separations or isopycnic separations

gradient pump. According to the operation sequence, it can be divided into over-layering and underlayering method.

Overlayering Method

Overlayering method is the method to prepare gradients simply by adding solutions from bottom to up, starting with the highest concentrated solution first. After adding the bottom layer, the subsequent solutions should be added as per the following instructions: (1) The centrifugal tube should be kept upright, and the solutions should be syringed enough slowly along the tube wall to the lower part. (2) The long needle tip can be bent into "L" shape, and the upper surface of the lower layer is covered by solution with low density. (3) A gradient instrument, such as DCF-U-type gradient instrument and can be used to track the liquid concentration.

Typically, for the preparation of a three-layer discontinuous gradient with the density of each layer of 1.2, 1.4, 1.6, and 1.8 g/cm^3, the solution with greatest density (1.8 g/cm^3) should be added to the bottom of the centrifuge tube according to the required volume, and then the solutions should be added with 1.6, 1.4, 1.2, and 1.2 g/cm^3 on the top step by step (Fig. 3.2).

Underlayering Method

The underlayering method is the method to prepare gradients by underlayering the required solutions one under the other, starting with the lightest solution first. A long and thin tube or needle can be inserted into the bottom of a centrifuge tube,

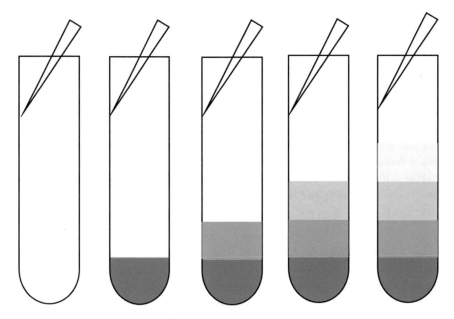

Fig. 3.2 Scheme of preparing density gradient via upper-spread method

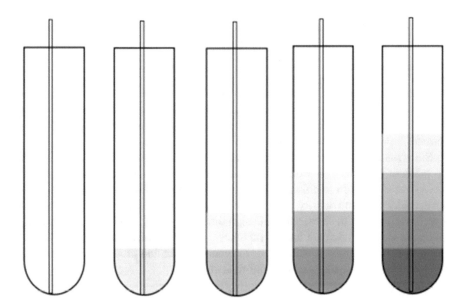

Fig. 3.3 Scheme of preparing density gradient via bottom-spread method

as shown in Fig. 3.3. The preparation process should be very careful, and the needle cannot be shaken and has to be maintained vertically, so as not to have a large impact on each layer.

It is worth noting that bubbles should be avoided during the gradient preparation. Too many bubbles will inevitably interfere the interfaces between the layers. Further, after layering the step gradient, it is necessary to check that there are sharp, refractive interfaces between each of the solutions to ensure that the gradient has been made correctly.

Once prepared, the step gradient should be used immediately; otherwise, it would probably transform into continuous gradient over several hours owing to solvent diffusion. If required, the sharp steps can be "rounded" by leaving the gradient standing vertically for about 2 h.

3.2.2 Continuous Gradient

The density of continuous gradient varies smoothly and continuously, which are usually linear or exponential, and has been advocated for particular separations. Besides, gradients with more complex shapes are only used for centrifugal separations in zonal rotors.

The shapes of the continuous density gradient can be designed as linear, concave, or convex curve type (Fig. 3.1) according to the characteristics and requirements of samples. Among them, the linear gradient is the most common instance which could be prepared in the following strategies.

Diffusion Method

A linear gradient placed for a period of time can be formed with diffusion. This method starts with preparing a step gradient whose density interval should be about 1–2 cm thickness of each layer of liquid and the relative linear gradient ought to be formed through free diffusion in a vibration-free condition.

Since the time required for diffusion varies according to the diffusion coefficient of gradient materials, it would be better to select the gradient media of low molecular weight. In addition, the difference of temperature, concentration, viscosity, thickness, and density of each layer has certain influence on the time to form a gradient. In order to shorten the diffusion time, a wire ring can be used by moving along the wall up and down for several times [6]. The diffusion method can stabilize the sample zone, which is suitable for the coarse separation process. However, it is not strict for quantitative analysis and time-consuming.

Gradiometer Method

A linear gradient can be easily prepared using an apparatus, as shown in Fig. 3.4. It is composed of two chambers with an outlet and a stirrer in the mixing chamber while the other chamber is the reservoir. The gradient is prepared with two different

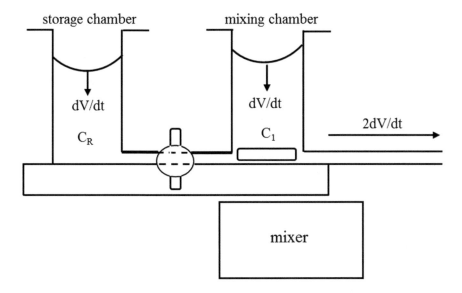

Fig. 3.4 Scheme of a simple two-cylinder gradient apparatus [6]

concentrations of solution, a high-density solution (the heavy solution) and a solution of low density (the light solution). The equal volume of light solution and heavy solution was added into two chambers, and a solution flows at a rate of dV/dt (V is the volume, t is the time) from one room to the other one. The mixed liquid flows out at a rate $2dV/dt$.

The concentration of the solution from the mixing chamber can be calculated by using Eq. 3.1.

$$C_t = C_M + (C_R - C_M)V_t/V_1 \qquad (3.1)$$

where C_t is the concentration of the liquid outflows from the mixing chamber at time "t", C_R is the concentration of the liquid in the storage chamber, C_M is the initial concentration of the liquid in the mixing room, V_t is the volume of the liquid flows out at the time "t", and V_1 is the original volume of the liquid in each room.

If C_R equals the concentration of the gradient light edge, V_0 becomes the final volume of the gradient, so that the bigger C_M is, the larger step the gradient has.

The formation of linear gradient is assumed simple, but, in fact, it is rather difficult to obtain real linear gradient. In order to gain a satisfactory linear gradient, the following conditions must be achieved: The two rooms must have the same geometric shape, and their respective liquid volumes should be the same at any time as well as reduce at equal rate, so the corresponding device must be utilized to control the flow rate.

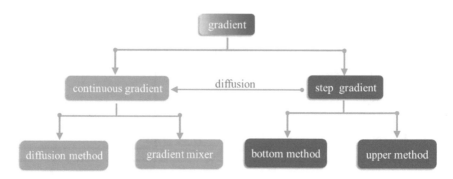

Fig. 3.5 Schematic of the gradient types and relevant preparation methods

As stated previously, the density gradient can be prepared by several methods relying on the type of gradient required and individual preferences. Figure 3.5 schematically presents the category of gradients and preparation methods.

Based on the above analysis, the advantages of discontinuous (step) gradient can be summed up as: (1) simply made without special equipment; (2) high efficiency, 10 min is enough to finish separations in most cases; (3) high repeatability, the gradient solutions remain stable for at least 1 h; (4) adjustable, the density range could be further narrowed down according to practical preseparation results. Therefore, the step gradient is widely applied for NPs separations.

3.3 Choice of a Rotor Type

Centrifugal rotor is the core component of the centrifuge, and all systems in centrifuge are configured to ensure the rotor under proper conditions. Rotor not only affects the separation effect, but also it is the main stress component of the centrifuge, so the safe operation of the centrifuge is very important. This section mainly introduces the classification of the centrifuge rotor and rotor-type selection.

3.3.1 Classification of Centrifuge Rotors

According to the properties, the centrifugal rotor can be divided into two types: the analysis of the rotor and the preparation of the rotor. Here, we mainly discuss the classification of the rotors. Material, shape, volume, and utility of each rotor are not identical, what we commonly use are fixed-angle rotor, swing-out rotor, zonal rotor, and vertical rotor.

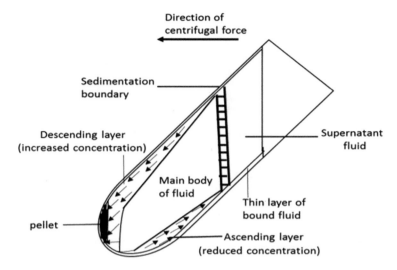

Fig. 3.6 Schematic diagram of a fixed-angle rotor

3.3.1.1 Fixed-Angle Rotors

Fixed-angle rotors (Fig. 3.6), as the name suggests, usually have several centrifugal tube cavities with a fixed angle. Angles between the center of the cavity axis and the axis of rotation range from 14 to 40°. The greater the angle, the better separation effect is.

Under the centrifugal force, particles will settle across the tube until hitting to centrifuge tube wall and then slide down to the bottom of the tube, which is called "wall effect" phenomenon.

The advantage here is that fixed-angle rotor has a large capacity, the center of gravity is low, the operation is balanced, and so fixed-angle rotor has a long lifetime.

The disadvantage is that the wall effect is easy to cause the settling particles to be disturbed by the sudden change of speed.

3.3.1.2 Swing-Out Rotors

Swing-out rotor (Fig. 3.7) usually has four or six free buckets (centrifugal casing); when the rotor is in static status, buckets are hung vertically. When the rotating speed reaches 200–800 rpm per minute, buckets will swing horizontally; this rotor is the best choice for density gradient zone centrifugation.

The advantage is that gradient material can be placed in the vertical centrifugal tube. Unlike the sediment samples in fixed-angle rotor, it is easy to take out the separated sample from the tube after the separation. While the disadvantage is that

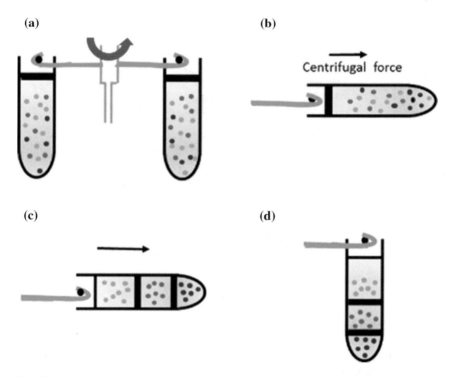

Fig. 3.7 Separations in swing-out rotors. The tubes are filled. Loaded into each bucket and attached to central body of the rotor, at rest the buckets of the rotor hang vertically. **a** As the rotor begins to move, the buckets move out so that they are perpendicular to the axis of rotation. **b** During centrifugation, the particles sediment down the tube. **c** When the rotor come to a stop. **d** The bucket return to a vertical position, and there is no reorientation of the liquid in the tubes

the sedimentation distance of particles is long, correspondingly the centrifugal time will be long.

3.3.1.3 Zonal Rotors

Zonal rotor is mainly composed of a rotor barrel and an unscrewed top cover Wherein, the rotor barrel is equipped with a cross-shaped clapboard device, which divides the rotor barrel into four or more chambers. The gradient liquid or sample liquid is pumped from the inlet pipe in the center of the rotor and is distributed around the rotor through a conduit inside the clapboard. The clapboard in the rotor can maintain the stability of the sample band and the gradient medium.

The sedimentation situation in zonal rotor differs from fixed-angle rotor or swing-out rotor. Under radial scattering effect of centrifugal force, particle settling distance will be unchanged, so the zonal rotor has a little "wall effect."

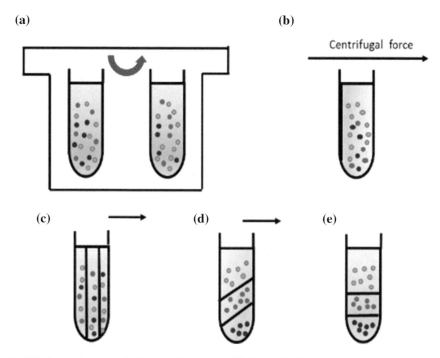

Fig. 3.8 Separation in vertical rotors. The tube is filled completely, sealed, and loaded into the pockets of the rotor (**a**). As the rotor begins to move, the liquid in the rotor reorientates through 90° (**b**). And the particles start to sediment across the tube (**c**). As the rotor decelerates below 1000 rpm, the bands reorientate (**d**). And as the rotor comes to a stop, this process is completed (**e**)

The advantage is that zonal rotor can avoid zone and settling particles becoming disorderly, which can also avoid disturbance of zone and settling particles. Furthermore, good separation effect, high speed, easy for gradient recovery, and having no influence on the resolution are the merits. In particular, this method can separate a large quantity of samples at once, about 60 ml. Compared to the chromatographic separation method, this method has great advantage in separation of sample with large capacity.

The disadvantage of zonal rotor is that the direct contact is inevitable between the sample and the medium, the corrosion resistance is high, and the operation procedure is complex.

3.3.1.4 Vertical Rotors

The centrifuge tubes of vertical rotor (Fig. 3.8) in the centrifugal process are kept to be vertical, that is the parallel orientation of the rotating shaft and the centrifuge tube. Compared to the fixed-angle rotor and swing-out rotor, the vertical rotor shortens the centrifugal time and at the same time improves the separation resolution.

Advantages are listed as following. The vertical rotor is mainly used for isopycnic gradient centrifugation. And because there is no wall effect (at least in the early stage of the gradient), it is also ideal for rate-zonal centrifugation. The difference between the minimum centrifugal acceleration and the maximum centrifugal acceleration of the vertical rotor is not very obvious, and all kinds of particles are in the strong centrifugal field, so the separation speed is fast and the resolution is high.

Disadvantages of vertical rotor are also illustrated here. Due to the redirection of the vertical rotor, the sample band in the centrifuge is wider in the latter part of the centrifuge than in the other rotor samples. Another problem is that the material that precipitates or floats on the gradient will be distributed over the length of the tube, and it can contaminate the supernatant when the centrifuge is complete.

3.3.2   The Selection Principle of the Rotors

Fixed-angle rotor has a relatively short distance for the centrifugal settling and recovery. That is to say, gradient curve flattens and the extension of the distance are conducive to the recycle of particles after centrifugation, so the settling distance of vertical rotor extends greater than the fixed-angle rotor; so vertical rotor is more conducive to isopycnic separation. On the contrary, granular zone width in the swing-out rotor is the narrowest, and the width of vertical rotor is the widest. It must be pointed out to in order to use the fixed-angle rotor for density gradient centrifugation, the separation of sample should not be too much, centrifugal speed is unfavorable and exorbitant, separation time should not be too long, otherwise, the precipitation will be attached to the wall, and the following causes pollution and influences the separation effect.

Swing-out rotor has the longest particle settling path, from meniscus to the bottom of the centrifuge tube. It also has merits including a wide range of density change toward to distance, easy to keep a wide gap between particles zones, can effectively prevent the formation of mixture, suitable for the separation and purification of larger virus particles and subcellular particles, but the swing-out rotor forming gradient costs a long time and, as a result, low efficiency of centrifugal.

Zonal rotor has the largest volume, which can separate about 60 mL samples once, so it is convenient for large quantity separation. Moreover, this method can avoid disorder of zone and settling particles and ensure good separation efficiency.

Vertical rotor has the shortest settling distance, centrifugal time, and the highest centrifugal efficiency, which is suitable for the extraction and purification of small particles. Due to the reorientation of liquid gradient medium in the centrifugal process and the centrifugal stop, the sample which is easy to produce "attached wall" is not suitable. And due to the circulation mixing and certain requirements for the concentration of the gradient, the viscosity of the medium, the use of fixed-angle rotor has a certain limit in the density gradient centrifugation.

3.3.3 The Relationship Between the Type of the Rotor and Centrifugal Force

In the ultracentrifuge, the specific rotor type corresponds to a specific speed range, and the specific speed relates to centrifugal force. Therefore, in different systems, the choice of rotor type is actually the choice of centrifugal force. It is well known that the sedimentation speed of the particles goes up as the centrifugal force increases. Thus, more rapid separation can be achieved by increasing centrifugation speed. Here, we briefly introduce some of the concepts of centrifugal force and the relationship between sedimentation speed and centrifugal force.

3.3.3.1 Centrifugal Force

At a certain angular velocity, any object that has a circular motion is subjected to an outward centrifugal force. The size of centrifugal force (Fc) is equal to the product of the centrifugal acceleration ($\omega^2 X$) and the particle mass (m), that is:

$$Fc = m\omega^2 X \tag{3.2}$$

The ω is the rotating angular, unit in radians/second; X is the distance from the center of the rotation to the particles; use cm as a unit; m is the mass in (g).

It can be seen that the size of the centrifugal force is proportional to the square of the rotational speed and the radius of rotation. Under certain conditions, the more far away from the axis of the particle, the greater the centrifugal force is. In the process of centrifugation, as the particles move in the centrifuge tube, the centrifugal force is also changed.

In practical work, the applied centrifugal force data is the mean data of centrifugal force which distributed throughout the centrifugal tube. That is, particles are affected by centrifugal force at the midpoint of the centrifugal solution.

3.3.3.2 Relative Centrifugal Force (RCF)

Commonly used "relative centrifugal force" or "digital * G" indicates that the centrifugal force, as long as the RCF value is constant, a sample can get the same centrifugal results under different centrifuge.

RCF is the actual centrifugal field that is converted to multiples of acceleration of gravity. That is:

Fig. 3.9 Conversion
of speed and relative
centrifugal force

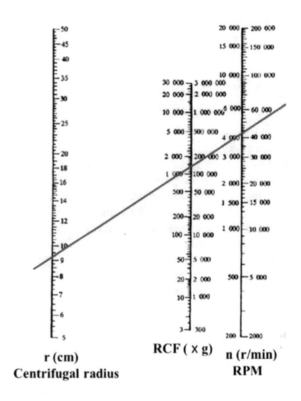

r (cm)
Centrifugal radius
    RCF (× g)   n (r/min)
                              RPM

$$\text{RCF} = \frac{F_{\text{Centrifugal force}}}{F_{\text{Gravity}}} = \frac{m\omega^2 X}{mg} = \frac{\omega^2 X}{g} = \frac{2\pi n X}{60/980} = 1.118 \times 10^5 \times X \times n \quad (3.3)$$

The X is the distance from the center of the rotation to the particles; use cm as a
unit and n on behalf of the centrifuge speed (rpm/min).

3.3.3.3 The Selection of Speed (N) and Relative Centrifugal Force (RCF)

As shown in Fig. 3.9, the "r" is the distance from the middle of the centrifuge tube
to the center of the centrifuge shaft center (cm) and "n" on behalf of the centrifuge
speed (rpm/min).

If the centrifuge speed needs to be converted to centrifugal force, firstly a known
radius should be picked up in the "r" scale and the centrifugal speed should be
taken in the "n" scale and extends to the RCF scale; intersection of lines is the
corresponding value of the centrifugal force. If RCF is known, centrifuge speed can
also be got.

In order to make specific rotor centrifugal conditions and get the same separation effect on another type of rotor, the relationship between the relative centrifugal force and the centrifugal time should be considered.

$$T_1 RCF_1 = T_2 RCF_2 \tag{3.4}$$

3.4 Choice of an Appropriate Separation Speed and Time

As discussed in the introduction of Sect. 3.2.1, we know that the density gradient centrifugation mainly includes three ways: differential centrifugation, isopycnic centrifugation, and rate-zonal centrifugation. With different choices of various methods, the calculation of the centrifugal time is also different.

In general, due to the changing shape of the gradient, the separation time of rate-zonal centrifugation is no more than 2 h. In terms of isopycnic centrifugation, it takes at least 9 h for the samples to be separated thoroughly. There is no clear statement for the separation time between 2 and 9 h, and the exact time depends on the specific samples. The general rule is that with the speed increases, the less time-consuming.

3.4.1 Differential Centrifugation

For differential centrifugation, the centrifugal time is the time when a particle is completely settled to the bottom of the tube. It is different from the isopycnic centrifugation or rate-zonal centrifugation; the influencing factors on centrifugal time of differential centrifugation are not complex, and the settling time can be obtained through calculation.

K coefficient is a characteristic parameter to characterize the efficiency of rotor centrifugal separation. The K coefficient of the rotor is taken into account in the centrifugal distance and the maximum centrifugal acceleration, when $S_{20,w}$ is known (sedimentation coefficient of water at 20 °C); the K coefficient can be used to calculate the centrifugal time of the maximum speed.

$$t = \frac{K}{S_{20,W}} \tag{3.5}$$

Therefore, if the K coefficient of a rotor and $S_{20,w}$ value are known, then the centrifugal time can be calculated at the maximum speed.

Among them, the K coefficient is related to the centrifugal distance and rotational speed of the rotor.

$$K = \frac{(\ln r_2 - \ln r_1)}{\omega^2} \times \frac{10^{13}}{3600} \tag{3.6}$$

In the above formula, speed of revolution $\omega = 0.10472n$(r/min); r_1: the distance from the center of the rotating shaft to the liquid level of the sample (cm); r_2: the distance from the center of the rotating shaft to the bottom of the centrifuge tube (cm).

Practical Application

Chen and co-workers reported the isolation of Au nanoparticle dimers in high purity by exploiting the high density of CsCl solutions. There is a typical setup of differential centrifugation, where 62% of CsCl and 11% of Au nanoparticle clusters with polystyrene-block-poly (acrylic acid) (AuNPn@PSPAA) in water were layered from bottom to top. To partition the AuNPn@PSPAA, a concentrated nanoparticle solution was carefully overlayered atop the 11 + 62% CsCl gradient, and the solution was then centrifuged in a desktop microcentrifuge at 8500 rpm (5800 g) for 20 min. The resulting solution showed two distinct bands of red and purple color, separated by a gap. The sample was extracted, purified to remove the excess CsCl, and then got the Au nanoparticle dimers with high purity [28].

3.4.2 Isopycnic Centrifugation

For isopycnic centrifugation, the time of centrifugation is equivalent to the time when the particles reach the point of equal density completely.

In most of the equilibrium gradient centrifugation, what added to the centrifuge tube is the mixture that evenly mixed with the sample and gradient medium. As long as the centrifugal time is long enough, the centrifugal gradient medium would form equilibrium gradient under the action of centrifugal force; at the same time, samples exist in each settling or floating, which close to their density, and finally the sample zone is formed. In the centrifuge, large particles reach the equilibrium position very quickly, and the smaller particles take a longer time. Therefore, the estimated time required to reach the equilibrium position of the particle is the standard time for small particles. Through the next calculation:

$$T = \frac{9.83 10^{13} \beta^0 (\rho_P - 1)}{N^4 r^2 S_{20,\,w}} \tag{3.7}$$

In the above formula, N is the actual speed, r/min; ρ_p stands for the buoyant density of samples; r is the distance from the center of rotation to the sample zone center, cm; β^0 see β^0 coefficient table; $S_{20,\,w}$ is the sedimentation coefficient for the sample at 20 °C, water as the medium.

Practical Application

Christine M. Nolan et al. separated the PEG cross-linked poly(N-isopropylacrylamide) microgel particles using isopycnic centrifugation in sucrose medium. The sucrose density gradient equilibrium centrifugation studies were carried out using a Beckman ultracentrifuge with 10-mL ultraclear centrifuge tubes. A sucrose density gradient was made by carefully layering 25, 20, 15, and 10% sucrose solutions (2.0 mL each) within the centrifuge tubes and then depositing 0.03 mL of the PEG cross-linked pNIPAm microgel dispersion on top. The samples were then centrifuged at 26 °C and 25,000 rpm for 4 h to allow the particle bands to reach their equilibrium density zones. Actual densities of these particles were not calculated, but qualitative differences were obtained by comparing the distances travelled down the sucrose density gradient. Samples were prepared in triplicate and presented as an average value ±1 standard deviation [27].

3.4.3   Rate-Zonal Centrifugation

For rate-zonal centrifugation, centrifugal time is the time of the formation of a well-defined zone.

Unlike differential centrifugation, there are many factors that affect the separation effect of particles in rate-zonal centrifugation. Due to the big difference between $\rho_{Tsolv}, \eta_{Tsolv}$ and $\rho_{20,w}, \eta_{20,w}$, even in the top of the gradient, and the $\rho_{Tsolv}, \eta_{Tsolv}$ is continuous changing, the separation requirements for particle centrifugation speed and time are affected not only by the influence of particle size, but also by gradient density range, the gradient shape, and concentration. Therefore, it is very difficult to determine the centrifugal state of the rate-zonal centrifugation. The most commonly used are the following two methods of estimation.

3.4.3.1 $\omega^2 t$ Calculation Method

This approach is to divide gradient into a series of small radius of components along the direction of the centrifugal force field. With the average density (ρ_M) of each increment of radius and viscosity (η_M) to calculate the time that the particle through the incremental required, the total of centrifugal effect is computed as:

$$\omega^2 t = \frac{(\rho_p - \rho_M)}{S_{20,w}\eta_{20,w}} \sum_{r_1}^{r_2} \frac{\eta_M \Delta r}{(\rho_p - \rho_M)r} \tag{3.8}$$

In the above formula, ρ_M and η_M refer to the density and viscosity of the center of each gradient in radius; r is the radius of increment center; r_1 and r_2 refer to the radius of the distance of the particle zone at the beginning and end of the operation, respectively.

Table 3.2 K' coefficient of some rotors

Rotor	The highest speed(r/min)	K' coefficient			
		1.2 g/cm^3	1.4 g/cm^3	1.6 g/cm^3	1.8 g/cm^3
Beckman					
SW60Ti	60,000	160	135	128	124
SW41Ti	41,000	379	317	300	292
SW25.1	25,000	1035	867	822	799
Dupont/Sorvall					
AH627	27,000	816	683	646	629
TST60.4	60,000	301	252	239	232
TH41	41,000	420	353	334	325
TV-850	50,000	115	95		

The calculation process of the method is complex, so it is rarely used in actual work, but the $\omega^2 t$ calculation method with many applications is applied in centrifugal experiment.

3.4.3.2 K′ Coefficient Estimation Method

In the rate-zonal centrifugation, the K' coefficient is used to estimate the time of particle precipitation, and the equation is as follows:

$$t = \frac{K'}{S_{20,W}} \tag{3.9}$$

In the formula, t is the centrifugal time; $S_{20,\ w}$ refers to particle sedimentation coefficient in standard condition; K' for the rotor's clarification coefficient; its value depends on the gradient, temperature, and particle density.

Table 3.2 lists some of the K' coefficients of the rotor, which is a linear sucrose gradient of 5–20%, with a temperature of 5 °C. With the K coefficient of each circumstance is close to each other, the lower K' coefficient is, the higher separation efficiency is.

Practical Application

George M. Whitesides et al. created a new approach to size-dependent and shape-dependent separation of nanoparticles through rate zonal centrifugation using multiphase systems as separation media. They use this method to separate the reaction product (nanorods) and byproducts (nanospheres and bigger particles) of a synthesis of gold nanorods. After estimation, they found the RCF must be greater than 8,800 g for a 25-nm (diameter) gold nanosphere (dgold = 19.3 g/cm3) in order to overcome the effect of the interfacial surface energy. They chose to use 16,000 g to ensure that they were well beyond the limit and also to reduce the time required to separate the objects [28].

References

1. Sartory WK (1970) Fractional cleanout in a continuous-flow centrifuge. Sep Sci Technol 5 (2):137–143
2. Perardi TE, Leffler RAA, Anderson NG (1969) K-series centrifuges II. Performance of the K-II rotor. Anal Biochem 32(3):495–511
3. Prantner AM, Chen J, Murray CB et al (2012) Coating evaluation and purification of monodisperse, water-soluble, magnetic nanoparticles using sucrose density gradient ultra-centrifugation. Chem Mater 24(21):4008–4010
4. Sun X, Luo D, Liu J et al (2010) Monodisperse chemically modified graphene obtained by density gradient ultracentrifugal rate separation. ACS Nano 4(6):3381–3389
5. Speranskaya ES, Beloglazova NV, Abé S et al (2014) Hydrophilic, bright $CuInS_2$ quantum dots as Cd-free fluorescent labels in quantitative immunoassay. Langmuir 30(25):7567–7575
6. Pine GG, Yuan XL (2008) Modern separation science and technology series: centrifugal separation (M). Chemical Industry Press, Beijing
7. Kirkland J, Yau W, Doerner W, Grant J et al (1980) Sedimentation field flow fractionation of macromolecules and colloids. Anal Chem 52(12):1944–1954
8. Sun X, Zaric S, Daranciang D et al (2008) Optical properties of ultrashort semiconducting single-walled carbon nanotube capsules down to sub-10 nm. J Am Chem Soc 130(20):6551–6555
9. Green AA, Hersam MC (2007) Ultracentrifugation of single-walled nanotubes. Mater Today 10(12):59–60
10. Green AA, Duch MC, Hersam MC (2009) Isolation of single-walled carbon nanotube enantiomers by density differentiation. Nano Res 2(1):69–77
11. Green AA, Hersam MC (2009) Solution phase production of graphene with controlled thickness via density differentiation. Nano Lett 9(12):4031–4036
12. Ghosh S, Bachilo SM, Weisman RB (2010) Advanced sorting of single-walled carbon nanotubes by nonlinear density-gradient ultracentrifugation. Nat Nanotechnol 5(6):443–450
13. Liu J, Hersam MC (2010) Recent developments in carbon nanotube sorting and selective growth. MRS Bull 35(4):315–321
14. Deng L, Wang X, Kuang Y et al (2015) Separation of non-sentimental carbon dots using "hydrophilicity gradient ultracentrifugation" for photoluminescence investigation. Nano Res 8 (9):2810–2821
15. Zhang C, Luo L, Luo J et al (2012) A process-analysis microsystem based on density gradient centrifugation and its application in the study of the galvanic replacement mechanism of Ag nanoplates with $HAuCl_4$. Chem Commun 48(58):7241–7243
16. Bonaccorso F, Zerbetto M, Ferrari AC et al (2013) Sorting nanoparticles by centrifugal fields in clean media. J Phys Chem C 117(25):13217–13229
17. Chang Z, Wu C, Song S et al (2013) Synthesis mechanism study of layered double hydroxides based on nanoseparation. Inorg Chem 52(15):8694–8698
18. Chen G, Wang Y, Tan LH et al (2009) High-purity separation of gold nanoparticle dimers and trimers. J Am Chem Soc 131(12):4218–4219
19. Wang Y, Chen G, Yang M et al (2010) A systems approach towards the stoichiometry-controlled hetero-assembly of nanoparticles. Nat Commun 1:87
20. Yang M, Chen T, Lau WS et al (2009) Development of polymer-encapsulated metal nanoparticles as surface-enhanced Raman scattering probes. Small 5(2):198–202
21. Bai L, Ma X, Liu J et al (2010) Rapid separation and purification of nanoparticles in organic density gradients. J Am Chem Soc 132(7):2333–2337
22. Ma X, Kuang Y, Bai L et al (2011) Experimental and mathematical modeling studies of the separation of zinc blende and wurtzite phases of CdS nanorods by density gradient ultracentrifugation. ACS Nano 5(4):3242–3249
23. Song S, Kuang Y, Liu J et al (2013) Separation and phase transition investigation of Yb 3 +/ Er 3 + co-doped NaYF4 nanoparticles. Dalton T 42(37):13315–13318

24. Stürzl N, Hennrich F, Lebedkin S et al (2009) Near monochiral single-walled carbon nanotube dispersions in organic solvents. J Phys Chem C 113(33):14628–14632
25. Mastronardi ML, Hennrich F, Henderson EJ et al (2011) Preparation of monodisperse silicon nanocrystals using density gradient ultracentrifugation. J Am Chem Soc 133(31):11928–11931
26. Guglielmi L, Battu S, Le Bert M, Faucher JL, Cardot PJP, Denizot Y (2004) Mouse embryonic stem cell sorting for the generation of transgenic mice by sedimentation field-flow fractionation. Anal Chem 76(6):1580–1585
27. Nolan CM, Reyes CD, Debord JD, García AJ, Lyon LA (2005) Phase transition behavior, protein adsorption, and cell adhesion resistance of poly(ethylene glycol) cross-linked microgel particles. Biomacromol 6(4):2032
28. Akbulut O, Mace CR, Martinez RV, Kumar AA, Nie Z, Patton MR, Whitesides GM (2012) Separation of nanoparticles in aqueous multiphase systems through centrifugation. Nano Lett 12(8):4060–4064

Chapter 4
Particle Sedimentation Behaviors in a Density Gradient

Pengsong Li

Abstract Density gradient centrifugation, as an efficient separation method, is widely used in the purification of nanomaterials including zero, one-, and two-dimensional nanomaterials, such as FeCo@C nanoparticles, gold nanoparticles, gold nanobar, graphene, carbon nanotubes, hydrotalcite, zeolite nanometer sheet (the examples can be found in Chap. 5). Each system needs separation parameter optimization, which comes from tremendous research experiments. When particles are put on the top of density gradient medium, they will have a definite settling rate under centrifugal force (F_c) [1], which is influenced by their net density, size, and shape. In a sufficiently intense centrifugal field, the particle motion held quietly free from gravity and vibration [2]. This is the principle of the density gradient ultracentrifuge. Based on the above principle, we discussed the particle sedimentation behaviors and built the kinetic equation in a density gradient media. The kinetic equation could apply to zero, one-, and two-dimensional nanomaterials, within its variation form accordingly. We found that the separation parameters could be optimized based on the kinetic equation. A MATLAB program was further developed to simulate and optimize the separation parameters. The calculated best parameters could be deployed in practice to separate given nanoparticles successfully.

Keywords Density gradient centrifugation · Sedimentation mechanism
Sedimentation kinetics · Mathematical model · Separation parameters
Optimization calculation

4.1 Sedimentation Mechanism of a Nanoparticle in a Centrifugal Field

In the centrifugal system, the driving force of the particle movement is the centrifugal force (F_c).

$$F_c = mG \tag{4.1}$$

where m is the mass of the particle (the unit is g). G is the centrifugal acceleration.

© The Author(s) 2018
X. Sun et al., *Nanoseparation Using Density Gradient Ultracentrifugation*,
SpringerBriefs in Molecular Science, https://doi.org/10.1007/978-981-10-5190-6_4

$$G = \omega^2 x \qquad (4.2)$$

where ω is the angular velocity of the rotor, the unit is rad/s, x is the distance from the nanoparticle to the rotation center, and the unit is cm.

So

$$F_c = m\omega^2 x \qquad (4.3)$$

The strength of the centrifugal force field can be described with the relative centrifugal field (RCF), namely the times of the gravitational acceleration (g).

$$RCF = \frac{G}{g} = \frac{\omega^2 x}{g} \qquad (4.4)$$

where the centrifugal angle is ω, and the distance from the particle to the rotation center is x. RCF can show the strength of the centrifugal force. For example, the rotor (P80AT) of Hitachi ultracentrifuge (CP80MX) can provide maximum RCF 615000 g with 80000 r/min.

One circle of the rotor is 2π radian. So, the rotational speed (n) of rotor can be described as follows:

$$n = \omega/2\pi \qquad (4.5)$$

Rotational speed is an important factor of separation, providing a specific centrifugal field in a given centrifuge [3]. Hence, appropriate centrifuge should be chosen to separate nanomaterials according to the centrifugal rotational speed.

The characteristic of particle movement in liquid medium not only depends on the centrifugal force but also rely on density, size, and shape of the particle and the density and viscosity (reverse viscous resistance (Reverse Friction)) of the liquid medium. The gravity and intermolecular force can be usually ignored because the centrifugal field is two orders of magnitude higher than the acceleration of gravity in general. Force analysis of the particle in the separation process is shown in the following figure (Fig. 4.1).

So, the dynamics equation of the particle in the centrifugal process can be described by the following differential equation:

$$m\frac{d^2 x}{dt^2} = F_c - F_b - F_f \qquad (4.6)$$

where t is the centrifugal time in units of s. F_b is the buoyancy, while F_f is the viscous resistance.

In different separation systems, the particle movement can be divided into three stages: The first one is an accelerative process with alterable positive acceleration; the second one is uniform motion without acceleration; and the last one is a

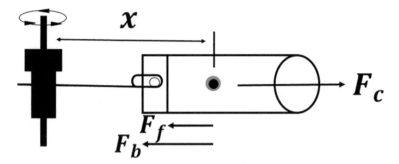

Fig. 4.1 Force analysis of the particle in a centrifugal field. Reprinted with permission from ref. [9]. Copyright 2016, American Chemical Society

decelerated stage with an alterable negative acceleration. When the particle achieves uniform motion [4], the force equilibrates ($d^2x/dt^2 = 0$) [1].

$$F_c = F_b + F_f \tag{4.7}$$

When the distance from the particle to the rotation center is x, and the mass of the particle is m, the centrifugal force can be calculated by Eq. (4.3).

According to Archimedes' principle, the buoyancy of the particle in the liquid medium is the weight of the displaced liquid.

$$F_b = V_p \rho_m g \tag{4.8}$$

where V_p is the volume of the particle, ρ_m is the density of the displaced liquid.

The buoyancy is proportional to particle volume but independent of particle shape. The volume of a particle is equal to the mass dividing a given density. So, Eq. (4.8) can be rewritten as:

$$F_b = \frac{m}{\rho_p} \rho_m g \tag{4.9}$$

This equation is set up in the gravitational field. While if a particle in the centrifugal field, the buoyancy can be considered as RCF times.

$$F_b = \frac{m}{\rho_p} \rho_m \omega^2 x \tag{4.10}$$

According to Stokes' Law, for a spherical rigid particle with a radius (r) which do not dissolve in the liquid medium, when the particle has a speed (dx/dt) under the centrifugal force field, it will be affected by the reverse viscous resistance (F_f).

$$F_f = 6\pi\eta r \times \frac{dx}{dt} \tag{4.11}$$

where η is the viscous coefficient of the liquid medium, r is the radius of the particle.

For non-spherical particles, they have much larger viscous resistance, and the frictional coefficient f is different from spherical particles f_0 (expression of $6\pi\eta r$). The relationship between f and f_0 can be described as $f = \theta f_0$, where the θ is the frictional ratio, and the θ value of other figurate nanostructures is usually in the range from 1 to 2.3, as shown in Fig. 4.2.

To accommodate particles of other shapes, one may apply the frictional ratio θ to Eq. (4.11),

$$F_f = 6\pi\eta r^* \times \frac{dx}{dt} \times \theta \tag{4.12}$$

where r^* is the radius of a sphere whose volume (V^*) is equal to that of the nanoparticles.

$$r^* = \sqrt[3]{\frac{3V^*}{4\pi}} \tag{4.13}$$

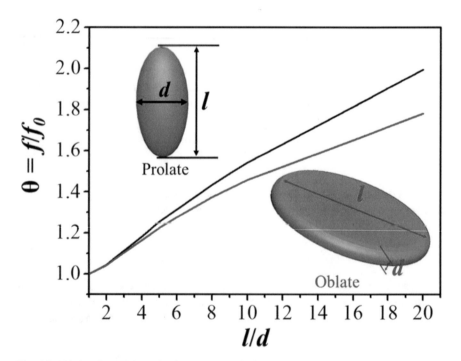

Fig. 4.2 Frictional coefficients for figurate non-spherical nanostructures [5]

For example, for specific rod-shaped CdS nanoparticles, the θ can be expressed as follows [1]:

$$\theta = 0.05213(L/D) + 0.954$$

where L is the length of the nanorod, D is the diameter of the nanorod.

Taking F_c, F_b, and F_f into Eq. (4.7), and using the product of volume and density to replace the mass, the sedimentation rate can describe as Eq. (4.14)

$$\frac{dx}{dt} = \frac{2r^{*2}(\rho_p - \rho_m)}{9\eta\theta} \omega^2 x \tag{4.14}$$

From the sedimentation rate formula, we should note:

1. r^* is the radius of a sphere whose volume (V^*) is equal to that of the nanoparticles (Eq. 4.14).
2. ρ_p is the net density of the particle, and ρ_p is the density of gradient media.
3. Theta (θ) is the frictional ratio.
4. For spherical particles: $\theta = 1$, $r^* = r$, r is the particle radius.

Some deductions can be drawn from above sedimentation rate formula:

1. The sedimentation rate of a particle is proportional to the square of the external particle diameter.
2. The sedimentation rate of a particle is proportional to the difference between particle density and medium density. When the difference is zero, the particle sedimentation will stop.
3. When the liquid medium viscosity increases, the particle sedimentation rate declines in proportion.
4. When the centrifugal field increases, the particle sedimentation rate increases in proportion to the centrifugal field.

In a given centrifuge system, the density and viscosity of the liquid medium are known quantities. For a certain particle, the r^*, ρ_p, η, and θ are also known quantity. So $2r^{*2}(\rho_p - \rho_m)/9\eta\theta$ can be defined as sedimentation coefficient.
That is

$$s = \frac{2r^{*2}(\rho_p - \rho_m)}{9\eta\theta} \tag{4.15}$$

and

$$\frac{dx}{dt} = s\omega^2 x \tag{4.16}$$

For convenience, we mark the sedimentation coefficient as follows.

$$s = \frac{dx/dt}{\omega^2 x} \tag{4.17}$$

On the other hand, from Eq. (4.18),

$$F_c = F_b + F_f \tag{4.18}$$

where $F_c = m\omega^2 x$, $F_b = m_{diss}\omega^2 x$, $F_f = fv$ and $m_{diss} = m\,\rho_{media}/\rho_{particle}$, v is the speed of particle, f is the fractional coefficient of particle.

Sedimentation coefficient [69] (s) can also be defined as Eq. (4.19)

$$s = dx/dt/\omega^2 x = m[1 - (\rho_{media}/\rho_{particle})]/f \tag{4.19}$$

Thus, there are mainly three factors can affect the sedimentation coefficient (s); (1) Effect of mass (m); greater the mass of particle, greater the sedimentation coefficient (s), the particle with higher mass travels down the centrifuge tube rapidly. (2) Effect of shape of particle (f = fractional coefficient of particle); more spherical particle moves with high sedimentation speed because more spherical particle has lower fractional coefficient value. (3) Effect of ($\rho_{media}/\rho_{particle}$) value; generally, ($\rho_{media}/\rho_{particle}$) value decides the sign of sedimentation coefficient (s) and particles settling orientations during the centrifugation; (a) when ($\rho_{media}/\rho_{particle}$) = 1, the value of sedimentation coefficient (s) is equal to zero, particles locate in the certain position that mean $\rho_{media} = \rho_{particle}$, (b) when ($\rho_{media}/\rho_{particle}$) < 1, the value of sedimentation coefficient (s) is greater than zero, particles settle along the direction of centrifugal force that mean $\rho_{media} < \rho_{particle}$, and (c) when ($\rho_{media}/\rho_{particle}$) > 1, the value of sedimentation coefficient (s) is less than zero, the particles float against the direction of centrifugal force that mean $\rho_{media} > \rho_{particle}$.

In physics, sedimentation means the sedimentation velocity under unit centrifugal force. If we take the experimental value into Eq. (4.17), the sedimentation coefficient (s) can be calculated. When the time is t_1, the position of the particle is x_1, and when the time is t_2, the position of the particle is x_2. The equation can be rewritten as shown as follows.

$$sdt = \frac{1}{\omega^2} \times \frac{dx}{x}$$

Integral in the above range:

$$s \int_{t_1}^{t_2} dt = \frac{1}{\omega^2} \int_{x_1}^{x_2} \frac{dx}{x}$$

Namely,

$$s = \frac{\ln(x_2/x_1)}{\omega^2(t_2 - t_1)} = 2.303 \frac{\lg(x_2/x_1)}{\omega^2(t_2 - t_1)} \quad (4.20)$$

Because of

$$\omega = \frac{2\pi n}{60} = 0.105n$$

We can get the relation between the sedimentation coefficient and the rotational speed (n):

$$s = \frac{2.1 \times 10^2 \lg(x_2/x_1)}{n^2(t_2 - t_1)} \quad (4.21)$$

where the unit of t_1 and t_2 is s, the unit of n is r/min.

We can choose centrifuges with different rotating speed according to the different sedimentation coefficients. In a certain separation system, we can estimate the time of centrifugal separation by Eq. (4.21).

4.2 Mathematical Description of Particle Sedimentation Kinetics

In the whole process of density gradient centrifugation, we can use Eq. 4.6 to describe the particle movement. In the liquid density gradient medium, the particle will have a solvation layer on the surface [6]. For ideal spherical particles with core density (ρ_c), radius (r), and solvation shell thickness (h) and the solvation shell density (ρ_h) (Fig. 4.3), the net density (ρ_p) can be estimated according to the following Equation,

$$\rho_p = \rho_h + (\rho_c - \rho_h)r^3/(r+h)^3 \quad (4.22)$$

It can be deduced from the above formula that the net density of a colloidal system would increase as the core size increases with respect to the hydration shell thickness, and the particle density will be close to the core material density when the nanoparticle is large enough (i.e., $r \gg h$).

Similarly, for the cylindrical particles with core density (ρ_c), radius (r), length (L), and hydrated shell thickness (h) in Fig. 4.4, the net density (ρ_p) can be estimated as:

$$\rho_p = \rho_h + \frac{(\rho_c - \rho_h)r^2 L}{(r+h)^2(L+2h)} \quad (4.23)$$

Fig. 4.3 Model of a
hydrodynamic colloidal
spherical nanoparticle.
Reprinted with permission
from ref. [9]. Copyright 2016,
American Chemical Society

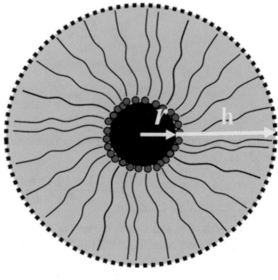

Fig. 4.4 Model of a
hydrodynamic colloidal
cylindrical nanoparticle

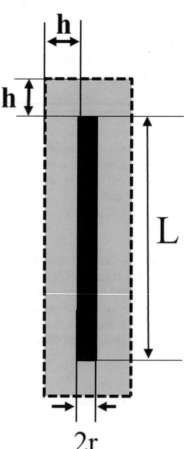

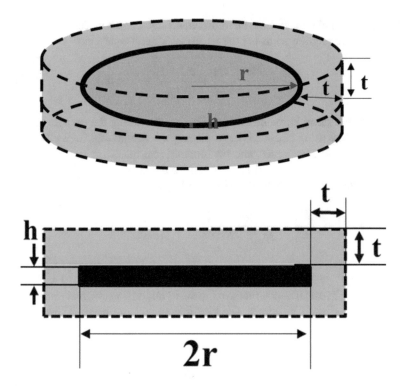

Fig. 4.5 A hydrodynamic colloidal nanosheets model

Similarly, for the two dimension nanosheet with core density (ρ_c), radius (r), thickness (h), and hydrated shell thickness (t) in Fig. 4.5, the net density (ρ_p) can be estimated as:

$$\rho_p = 1 + (\rho_c - 1)/[(1 + t/r)^2(1 + 2t/h)] \qquad (4.24)$$

The particles with different morphologies have different net densities accordingly, and all have a sedimentation tendency as layered on top of the linear density gradient, driven by centripetal force. The movement behavior is determined by the particle size, net density, centrifugal force, buoyancy, and viscous resistance.

Hence, for spherical particles, we can describe their movement during the centrifugal process using the following equation,

$$\begin{cases} m\frac{d^2x}{dt^2} = F_c - F_b - F_f \\ F_c = m\omega^2 x \\ F_b = \frac{m}{\rho_p}\rho_m\omega^2 x \\ F_f = 6\pi\eta r * \frac{dx}{dt}\theta \end{cases} \Rightarrow m\frac{d^2x}{dt^2} = m\omega^2 x - \frac{m}{\rho_p}\rho_m\omega^2 x - 6\pi\eta r * \frac{dx}{dt}\theta$$

The nanoparticle mass can be represented by apparent density, the above equation can be rearranged to get the universal kinetic equation:

$$\frac{d^2x}{dt^2} + \frac{9\eta\theta}{2\rho_p r*^2}\frac{dx}{dt} + \frac{\rho_m - \rho_p}{\rho_p}\omega^2 x = 0 \qquad (4.25)$$

All particle motion in the density gradient media during the centrifugal process can be described using Eq. 4.25. Let's discuss the formula in three different situations: applying to zero-, one-, and two-dimensional nanostructures.

Firstly, for zero-dimensional nanostructures: $\theta = 1$, $r* = r + h$, r is the particle radius, h is the hydrated shell thickness, the kinetic equation can be simplified as:

$$\frac{d^2x}{dt^2} + \frac{9\eta}{2\rho_p(r+h)^2}\frac{dx}{dt} + \frac{\rho_m - \rho_p}{\rho_p}\omega^2 x = 0 \qquad (4.26)$$

Secondly, for one-dimensional nanostructures: $r* = \sqrt[3]{3/4(r+h)^2(l+2h)}$, r is the radius of cylindrical nanostructure, l is the cylindrical nanostructure length, and h is the hydrated shell thickness. The kinetic equation can be simplified as:

$$\frac{d^2x}{dt^2} + [\frac{3}{4}(r+h)^2(l+2h)]^{-\frac{2}{3}}\frac{9\eta\theta}{2\rho_p}\frac{dx}{dt} + \frac{\rho_m - \rho_p}{\rho_p}\omega^2 x = 0 \qquad (4.27)$$

Lastly, for two-dimensional nanostructures: $r* = \sqrt[3]{3/4(r+t)^2(h+2t)}$, r is the nanosheets radius, h is the nanosheets thickness, t is the hydrated shell thickness, then the kinetic equation can be simplified as:

$$\frac{d^2x}{dt^2} + [\frac{3}{4}(r+t)^2(h+2t)]^{-\frac{2}{3}}\frac{9\eta\theta}{2\rho_p}\frac{dx}{dt} + \frac{\rho_m - \rho_p}{\rho_p}\omega^2 x = 0 \qquad (4.28)$$

The multifunctional, universal kinetic equation can be used in not only rate zonal separation but also isopycnic separation.

For rate zonal separation, which mainly uses the different settling rate to sort the nanoparticles, the max density of density gradient media is smaller than that of nanostructures ($\rho_m < \rho_p$) in general, and Eq. 4.25 can record the whole motion of nanoparticles.

For isopycnic separation, which mainly uses tiny differences of net density to sort the nanoparticles, the density of nanoparticles should locate in the range of the density of gradient media. When the nanoparticles reach the isopycnic state ($\rho_p = \rho_m$), the settling rate will become zero, and the nanoparticles will stay at that position even prolonging the time. The centrifugal time can also be calculated through Eq. 4.25, as well as the sedimentation coefficient (Eq. 4.20).

It can be deduced from the above formula (Eq. 4.25 for all nanostructures; Eq. 4.26 for zero-dimensional nanostructures; Eq. 4.27 for one-dimensional nanostructures; Eq. 4.28 for two-dimensional nanostructures) that the positions of the particles in the centrifugal tube after separation are determined by centrifugal time, centrifugal rotational speed, density gradient range, and medium viscosity, etc. Therefore, there will be a lot of factors affecting the separation effect.

4.3   The Influence of Separation Parameters

For optimized separation, various parameters should be considered: centrifugal rotational speed, centrifugal time, density gradient range, and medium viscosity and so on. An ideal separation should be described as: 1. The spatial distribution should be the longest, with the smallest and biggest NPs located at the top and bottom of the centrifugal tube. 2. The size distribution of the as separated particles along the centrifugal tube is linear. We will analyze the above factors in detail.

4.3.1   Influence of the Centrifugal Rotational Speed (ω)

At insufficient centrifugal rotational speed, the separated nanoparticles mainly distributed in the top half of the centrifuge tube, while excessive centrifugal rotational speed made the fractions concentrated in the bottom of the centrifuge tube. The two cases could not make the particles dispersed in the whole centrifugal tube. An appropriate centrifugal speed could fractionate particles very well, indicating better separation efficiency (Fig. 4.6).

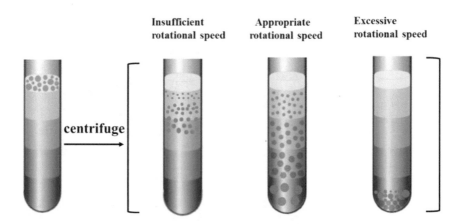

Fig. 4.6 Influence of centrifugal rotational speed at rate zonal separation

4.3.2 Influence of the Centrifugal Time (T)

The centrifugal time is another important factor in density gradient separation. For low-density materials, such as carbon nanotubes and graphene nanosheets with surfactant wrapping, their net densities are ~ 1.1 g/cm^3, and the gradient density should be tailored to cover particle net densities. Under centrifugal force, the low-density particles would move to and stay at the layers where the medium density equal to their net densities (isopycnic separation). No matter how long the centrifuge time is, the particles will not sediment down (Fig. 4.7a). While if the net densities of particles are higher than the highest density that the gradient media can reach (rate zonal separation), the particles will sediment through the gradient unless the external centrifugal force is removed [7]. Therefore, the particles with different sizes and net densities have different sedimentation velocities (Eq. 4.14), and they will stop at different locations after a given centrifugal time. When time is long

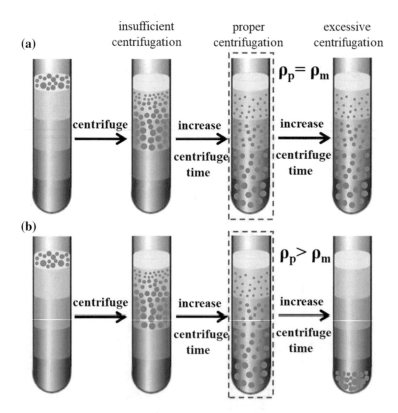

Fig. 4.7 Schematic illustration of typical isopycnic separation **a** and rate zonal separation **b**. The optimized separation states as shown in red boxes. Reprinted with permission from ref. [7]. Copyright 2014, Elsevier Inc. All rights reserved

enough, all the particles will sink to the bottom of the centrifuge tube (Fig. 4.7b). At a given centrifugal system, we should choose an appropriate centrifugal time for rate zonal separation.

4.3.3 Influence of the Density Gradient Range (ρ_m)

The choice of gradient media will also affect the separation efficiency. For a given nanoparticle, its net density is a constant. If the net density is relatively small, the isopycnic separation can be selected to sort the nanoparticles. It is worth noting that the range of the gradient media should cover the net density of nanoparticles. If the distribution of net densities is very narrow, a more precise density gradient should be chosen. For example, when the distribution of net densities is from 0.85 to 0.90 g/cm^3, we can use the density gradient with 0.84–0.91 g/cm^3, which can make the fractions fully distributed along the centrifugal tube; while the larger range (such as 0.7−1.0 g/cm^3) would not get such separation effect.

For high-density materials, such as metal, metallic oxide, and metallic selenide, the rate zonal separation is a better choice, and the density gradient should be also chosen accordingly. For instance, the net density of cadmium selenide quantum dots with the size range from 3 to 7 nm synthesized in 1-octadecene can be calculated by Eq. 4.22, which is ranged from 0.82 to 1.26 g/cm^3. For the separation of those CdSe quantum dots, cyclohexane/carbon tetrachloride density gradient can be chosen. Here, the density of cyclohexane is 0.78 g/cm^3 and carbon tetrachloride is 1.59 g/cm^3, so cyclohexane/carbon tetrachloride system can provide the density gradient range from 0.78 to 1.59 g/cm^3, which can be tuned to match the net density range of CdSe quantum dots. Moreover, if the density gradient range (e.g., 0.8–1.0 g/cm^3) leads the fractions enriching in bottom layers or the density gradient (e.g. 1.3–1.5 g/cm^3) leads the fractions accumulating in the upper layers of the centrifuge tube, the density gradient range can be further tuned to make the fractions fully distributed. Thus, whatever it is isopycnic separation or rate zonal separation, a suitable choice of density gradient range depends on the net density of nanoparticles.

To separate the particles synthesized in water should choose an aqueous gradient medium, such as cesium chloride aqueous solution, sodium chloride aqueous solution, and sucrose solution. While to separate the particles synthesized in oil should choose an organic gradient medium, such as cyclohexane/carbon tetrachloride, ethanol/ethylene glycol, and acetone/chloroform.

4.3.4 Influence of the Medium Viscosity (η)

Viscosity can affect the viscous resistance and also affect the stability of the density gradient. For the separation of particles with small net density, high viscosity liquid medium (e.g., ethylene glycol), will cost long time to get a good separation effect.

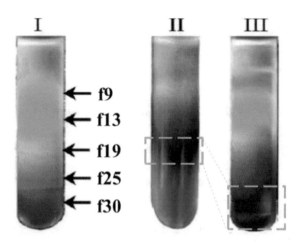

Fig. 4.8 Images of ultracentrifuge vessels containing cadmium selenide nanoparticles under UV irradiation at 365 nm: (Vessel I) polystyrene-free gradient, 60 min centrifugation at 50000 rpm, (Vessel II) polystyrene-containing gradient, 60 min centrifugation at 50000 rpm, and (Vessel III) polystyrene-containing gradient, 110 min centrifugation at 50000 rpm. Reprinted with permission from ref. [8]. Copyright 2010, American Chemical Society

In high viscosity density gradient medium system, larger centrifugal force is needed to make the particles move.

While in some conditions, the viscosity can be modified by introducing polymer without influencing density. When Bai et al. [8] used the cyclohexane/carbon tetrachloride density gradient to separate the cadmium selenide nanoparticles, and they studied the influence of viscosity on the separation effect by adding the polystyrene (PS) in the density gradient medium. As the introduction of PS into the organic gradient layers, it can significantly increase their viscosity; it should slow down the sedimentation of nanoparticles. As expected, PS-containing gradient (vessel II) showed a limited separation compared to PS-free gradient (vessel I), and only by applying longer centrifugation time, the separation can be completed (vessel III), and thus a finer separation can be achieved (Fig. 4.8). This work demonstrated the possibility to separate by tailoring media viscosity.

When using two or three kinds of liquid medium or a substance solution to prepare the density gradient, the viscosity of gradient mainly depends on the density. So we can build a relation between the viscosity and the density as follows:

$$\eta = \eta(\rho_m) \tag{4.29}$$

For sucrose as an example, we can find the following data in the *Chemical Property Manual: Organic Volume* (Table 4.1).

Table 4.1 Density and viscosity coefficient of different concentration sucrose solutions at 20 °C	Density of sucrose solution (g/cm^3)	Viscosity coefficient (mPa.s)
	1.2	1.957
	1.25	2.463
	1.3	3.208
	1.35	4.352
	1.4	6.21
	1.45	9.449

We analyze the experimental data of sucrose solution on density and viscosity coefficient and find the relationship between viscosity coefficient and density is similar to the quadratic polynomial relation. Therefore, we can assume that

$$\eta = \alpha \rho_m^2 + \beta \rho_m + \delta \tag{4.30}$$

Using the least square method to fit, we can obtain:

$$\alpha = 129.41; \beta = -314.45; \delta = 193.11$$

Experimental data fitting method can provide the relationships between the viscosity coefficient and the gradient medium in other separation systems.

Above analysis indicates that centrifugal parameters including centrifugal forces, centrifugation time, density range, and medium viscosity would influence the final efficiency of density gradient separation. Thus, in the succeeding section, we will discuss the mathematical optimization of density gradient separation.

4.4   Optimization Model for Best Separation Parameters

Based on above discussion, many factors (centrifugal rotational speed, centrifugal time medium viscosity, density gradient range, and hydrated shell thickness) would influence the final density gradient separation effect. Moreover, the factors are highly connected. For example, higher rotational speed needs shorter time, or lower rotational speed takes longer time. Therefore, a large number of control experiments are needed to explore the best separation parameters for a perfect separation result, while it will be time consuming and inefficiency.

To address this problem, we develop a mathematical optimization method to study the kinetic equation in the centrifugal process [9]. Although there are a lot of variables of the kinetic equation, and it is a nonlinear differential equation without exact solution, we can briefly consider the equation as the relation expression between nanoparticle size (r) and its position (x). After the simulation of these variables, we can get a good linear distribution between r and x, and find the optimized separation parameters.

In detail, we consider the centrifugal accelerating and de-accelerating process, and the following functions are used to describe angular speed.

$$\omega(t) = \begin{cases} \frac{\omega t}{T_0} & (0 < t \le T_0) \\ \omega & (T_0 < t \le T_0 + T_1) \\ \omega\left(\frac{T_0 + T_1 + T_2 - t}{T_2}\right) & (T_0 + T_1 < t \le T_0 + T_1 + T_2) \end{cases} \quad (4.31)$$

where ω is the stable angular speed of the centrifuge after accelerating process, T_0, T_1, and T_2 are the durations of accelerating, separating, and moderating processes.

In order to further simplify the optimization model, we assume that the ideal distribution of the fractions after density gradient separation is linear:

$$X(r) = ar + b \quad (4.32)$$

where X is the position of the particle with diameter of r, a and b are linear constants.

As illustrated above, the gradient interfaces would influence the sedimentation resistance a little, thus in order to get better separation and simplify the calculation, we assume that the ideal density gradient is linear:

$$\rho_m(x) = dx + c \quad (4.33)$$

where x is the distance between rotation center and the gradient with a density of ρ_m, c, and d are linear constants.

Therefore, based on above modeling, the location of a nanoparticle with a diameter of r is the function of separation time T_0, T_1, T_2, angular speed ω (i.e., centrifugal force), gradient ρ_m, and media viscosity η. The location of a particle with a diameter of r could be described as $X(T_0, T_1, T_2, \omega, d, c, a, b, \eta)$. The position X of a particle at the moment t can be calculated by the following differential equation as the initial states are given. There are two initial conditions when the separation starts: the particle's initial position is x_0 and initial velocity is 0. So we can get the following conditional equation:

$$\begin{cases} \frac{d^2x}{dt^2} + \frac{9\eta(\rho_m(x))}{2\rho_p(r+h)^2}\frac{dx}{dt} + \frac{\rho_m(x)-\rho_p(r)}{\rho_p(r)}\omega(t)^2 x = 0 \\ x(0) = x_0, x'(0) = 0 \end{cases} \quad (4.34)$$

Because the viscosity of the density gradient is related to the density of the gradient, the influence of the viscosity can be attributed to the density of the gradient. Then, the optimization could be carried out by using above X value. The objective function of the least square optimization model is then set up as:

$$G(T_0, T_1, T_2, w, d, c, a, b) = c_1^2[x_0 - (ar_0 + b)]^2 + c_2^2[x_m - (ar_m + b)]^2$$
$$+ \sum_{i=1}^{m-1} [X(T_0, T_1, T_2, w, d, c, a, b) - (ar_i + b)]^2$$

(4.35)

where c_1 and c_2 are the weighting factors, which contributing to the size distribution of the fractions. x_o and x_m are the upper and lower positions of the tube. For a setting parameter a, b, d, c_1, c_2, and c, when objective function G reach the minimal value, the gradient function $\rho_m(x)$ will give the ideal gradient.

A set of estimated separation parameters are given as initial values, by using the nonlinear least square method mentioned above, we could get the minimum value of the objective function G using a MATLAB program. After calculation, the MATLAB program output the optimized size distribution, which almost coincides with the ideal size distribution, as shown in the comparison chart in Fig. 4.9. It should be noted that the optimized distribution is just a calculated distribution other than the real distribution of particles. Nevertheless, the computer program will output the best separation parameters after calculation, and then we can use those separation parameters to separate our nanoparticles.

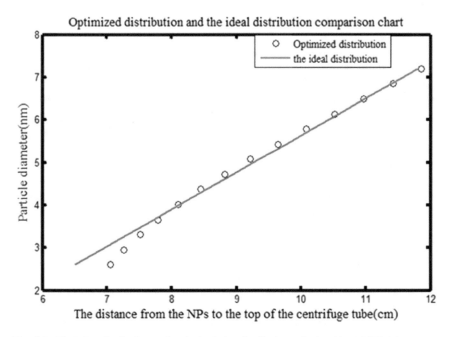

Fig. 4.9 Ideal size distribution and optimized size distribution calculated by MATLAB program using optimization model. Reprinted with permission from ref. [9]. Copyright 2016, American Chemical Society

For example, in the separation system of cadmium selenide nanoparticles using the cyclohexane and carbon tetrachloride density gradient, we can optimize the model to get the best separation parameters. According to the cadmium selenide nanoparticles with the size of 3–7 nm, the calculated separation conditions are 51012 rpm of centrifugal rotational speed, 1.65 h of centrifugal time within a 5–60% (cyclohexane/carbon tetrachloride) density range. Using those separation parameters, we can get ideal separation effect.

In addition, for the separation of other morphological nanostructures, a physical quantity of the shape factor can be introduced to correct the effect of morphology on the dynamics equation. As a result, the same optimization simulation method can output a series of suitable centrifugal parameters.

Appendix: MATLAB Program for the Computational Mathematical Optimization of Spherical Nanoparticles

% model assumptions:

%1. All the nanoparticles are sphere; if not, use morphology factor f to modify the model;

%2. The nanoparticles have a solvation layer;

%3. Using linear density gradient; if not, modify the gradient function;

%4. The ideal diameter distribution of nanoparticles is linear distribution; the objective function is G;

%5. Nanoparticles do not react with the medium;

%6. Optimization variables: linear acceleration time (T_0), the time of constant speed (T_1), linear deceleration time T_2, (the total time is $T = T_0 + T_1 + T_2$), the angular velocity in constant speed (omega), c and d are the coefficient of linear density gradient; a and b are the coefficient of the ideal linear distribution; the objective function is $G(T_0, T_1, T_2, \text{omega}, c, d, a, b) = c_1 \wedge 2 * (x_0 - (a * r_0 + b)) \wedge 2 + \text{sum}((x(r_j,T) - (a * r_j + b)) \wedge 2)(j = 0:m) + c_2 \wedge 2 * (x_m - (a * r_m + b)) \wedge 2$, ($c_1$ and c_2 are appropriate constant); among them, $r_j = r_0 + j * (r_m - r_0)/m$, r_0, r_m are the minimum and the maximum radius of particles, respectively; x_0 is the distance between the top of the centrifuge tube and the center of rotation, x_m is the distance between the bottom of the centrifuge tube and the center of rotation.

%7. The movement of NPs ($X(t)$) follows the following equation: $x'' + 9 * \text{ita} (p_m(x))/(2 * p_p(r) * r \wedge 2) * x' + (p_m(x) - p_p(r)) * \text{omega}(t)/p_p(r) * x = 0$; ita is the viscosity of the medium solution, the ideal density gradient: $p_m(x) = c + d * x$, the density of nanoparticle: $p_p(r)$; the angular velocity: $\text{omega}(t) = \text{omega} * t/T_0$ ($0 < t < T_0$); omega ($T_0 < t < T_0 + T_1$); $\text{omega} * (T_0 + T_1 + T_2 - t)/T_2$ ($T_0 + T_1 < t < T_0 + T_1 + T_2$);

% Using the Lsqnonlin in MATLAB to solve the optimization problem

global r x_0 x_m

$m = 10$; % the number of output dots

$x_0 = 6.5$; $x_m = 11.8$; $r_0 = 1.3$e-7; $r_m = 3.6$e-7; % the unit is centimeter, might be different for different rotors; x_0 and x_m is the distances between center of rotation and the top and bottom of the centrifuge tube, respectively; r_0 and r_m is the size of particles.

$h = (r_m-r_0) / m$; $r = r_0:h:r_m$;

$y_0 = [60,3600,60,4000,0.6,0.5,2e7,4]$; % the initial value of the optimization calculation

$l_b = [30,1200,30,1000,0,0,0,-1e8]$; % the minimum bounds of optimization variables

$u_b = [1000,100000,1000,10000,5,5,1e10,1e10]$; % the maximum bounds of optimization variables

options = optimset('LargeScale','on','Display','iter','TolX',1e-30, 'MaxIter', 200, 'MaxFunEvals', 5000, 'TolFun', 1e-10);

[y, resnorm] = lsqnonlin(@objfun, y_0, l_b, u_b,options) % optimization calculation

$a = y(7)$; $b = y(8)$;

arb = $a * r + b$; % the ideal linear distribution

Fm = objfun(y); xt = Fm(2:end-1)' + arb;

plot(xt, 2 * r,'o', arb, 2 * r) % output comparison chart

title('Optimized distribution and the ideal distribution comparison chart')

xlabel('The distance from the NPs to the top of the centrifuge tube(cm)')

ylabel('Particle diameter(cm)')

legend('Optimized distribution','the ideal distribution')

function F = objfun(y) % the objective function

global r x_0 x_m

$T_0 = y(1)$; $T_1 = y(2)$; $T_2 = y(3)$; omega = $y(4)$; $c = y(5)$; $d = y(6)$; $a = y(7)$;

$b = y(8)$;

$T = T_0 + T_1 + T_2$;

tspan = $[0,T]$;

xt = [];

m = length(r);

for i = 1:m

ri = $r(i)$; $[t,x]$ = ode15s(@odefun,tspan,$[x_0,0]$,[],y,ri);

xti = x(end,1);

xt = [xt,xti];

end

arb = $a * r + b$;

$c_1 = 10000$; $c_2 = 10000$;

$F = [c_1 * (x_0\text{-arb}(1)), \text{xt-arb}, c_2 * (x_m\text{-arb}(\text{end}))]$;

$F = F'$;

function xp = odefun(t,x,y,ri)

$p_c = 6$; $h = 2e-7$; $p_m = 0.9$; % p_c is the density of core, p_m is the density of shell, the unit is g/cm^3; h is the thickness of the shell, the unit is centimeter.

$p_p = p_m + (p_c-p_m) * (1-h/\text{ri}) \wedge 3$;

$T_0 = y(1)$; $T_1 = y(2)$; $T_2 = y(3)$;

$c = y(5)$; $d = y(6)$; omega = $y(4)$;

if $t >= 0$ & t $< T_0$
omegat = omega $* t / T_0$;
elseif $t >= T_0$ & $t <= T_0 + T_1$
omegat = omega;
elseif $t > T_0 + T_1$ & $t <= T_0 + T_1 + T_2$
omegat = omega $* (T_0 + T_1 + T_2-t)/T_2$;
else
omegat = 0;
end
$(c + d * x(1)) > 0.5$ & $(c + d * x(1)) < 2$; % the minimum and maximum bounds of the density gradient
ita = ita(p_m); % viscosity is relate with the density of liquid medium, the unit of viscosity is mPa.s
$x_p = [x(2);-9 * ita * x(2) / (2 * ri * ri * p_p) + (p_p-(c + d * x(1))) * omega * omega * x(1) / p_p]$; % the kinetic equation of spherical nanoparticles

References

1. Ma X, Kuang Y, Bai L, Chang Z, Wang F, Sun X, Evans DG (2011) Experimental and mathematical modeling studies of the separation of zinc blende and wurtzite phases of CdS nanorods by density gradient ultracentrifugation. ACS Nano 5(4):3242–3249
2. Mcbain JW (2002) Opaque or analytical ultracentrifuges. Chem Rev 2:289–302
3. Lin Y (2008) Centrifugal Separation(Modern separation science and technology books). Chemical Industry Press. 林元喜 (2008) 离心分离(现代分离科学与技术丛书). 化学工业出版社
4. Price CA (1982) Centrifugation in density gradients. Academic Press
5. Svedberg T, Pedersen KO (1940) The Ultracentrifuge. The Ultracentrifuge
6. Sun X, Tabakman SM, Seo WS, Zhang L, Zhang G, Sherlock S, Bai L, Dai H (2009) Separation of nanoparticles in a density gradient: FeCo@C and gold nanocrystals. Angew Chem Int Edit 121(5):957–960
7. Kuang Y, Song S, Huang J, Sun X (2015) Separation of colloidal two dimensional materials by density gradient ultracentrifugation. J Solid State Chem 224:120–126
8. Bai L, Ma X, Liu J, Sun X, Zhao D, Evans DG (2010) Rapid separation and purification of nanoparticles in organic density gradients. J Am Chem Soc 132(7):2333–2337
9. Li P, Huang J, Luo L, Kuang Y, Sun X (2016) Universal parameter optimization of density gradient ultracentrifugation using CdSe nanoparticles as tracing agents. Anal Chem 88 (17):8495

Chapter 5
Density Gradient Ultracentrifugation of Colloidal Nanostructures

Liang Luo, Qixian Xie and Yinglan Liu

Abstract According to the centrifugation theory, various factors, such as the media density (ρ_m), radius (r) and thickness (h) of nanostructures, and solvation shell thickness (t) in different media, will directly influence the particle behavior during the density gradient centrifugation process. Density gradient centrifugation has become a promising tool to purify nanomaterials, such as metal nanostructures, carbon materials (carbon nanotubes and graphene), non-metal nanostructures (e.g., rare-earth nanostructures and oxide nanostructures). For the practical separation, as demonstrated in previous chapters, on the basis of the theoretical analysis of the target nanostructures and the preliminary separation, one can optimize the centrifugation according to the comprehensive consideration. While after all, the optimization direction of nanoseparation should be mainly focused on the net density of nanostructures and media. In this chapter, we will discuss the separation examples according to the dimensional difference of colloidal nanostructures, including 0D, 1D, 2D nanostructures, and assemblies/clusters.

Keywords Zero dimensional · One dimensional · Two dimensional
Assemblies · Clusters

5.1 Separation of Zero-Dimensional Nanostructures

Quantitatively, a theoretical assay would be important for optimizing separation of zero-dimensional materials, and the kinetic equation of zero-dimensional nanostructures can be demonstrated as:

$$\frac{d^2x}{dt^2} + \frac{9\eta}{2\rho_p(r+h)^2}\frac{dx}{dt} + \frac{\rho_m - \rho_p}{\rho_p}\omega^2 x = 0 \qquad (5.1)$$

Zero-dimensional nanostructures with different size and shape could be separated with different rates. For the nanostructures with net density much higher than the media density ($\rho_p > \rho_m$), such as noble metal, they can usually be separated by rate-zonal separation with high efficiency. While if the net density of nanostructures

© The Author(s) 2018
X. Sun et al., *Nanoseparation Using Density Gradient Ultracentrifugation*,
SpringerBriefs in Molecular Science, https://doi.org/10.1007/978-981-10-5190-6_5

(ρ_p) is relative small or the density difference of fractions is very small, they would be separated by using isopycnic separation. Furthermore, the net density difference of nanostructures can be enlarged based on some strategies, such as changing the solvation shell thickness (t) by introducing surfactant, or increasing media viscosity by introducing polymers.

The separation of nanoparticles with high net density is usually simple. By using the aqueous multiphase systems (MuPSs) as media for rate-zonal centrifugation, Akbulut et al. [2] separated the gold nanoparticles from the nanorods with different sizes and shapes successfully (Fig. 5.1b). Similarly, Osman M. Bakr et al. [3] also used rate-zonal density gradient centrifugation method to separate diamond nanocrystals with diameters mainly <10 nm. While as the size decreases (lower than 5 nm), the net density is much smaller, and the density difference is also dramatically decreased due to the existence of solvation shell (see Eq. 4.20 in Chap. 4). Hence, for the separation of ~4 nm FeCo@C nanoparticles, Sun et al. [1] encapsulated the nanoparticles with polyethylene glycol (PEG), and enlarged the hydration shell thickness, namely, enlarged the density difference of nanoparticles. With the iodixanol/water solutions (20 + 30 + 40 + 60%) as density gradient media and rate-zonal separation method, FeCo@C nanocrystallines with coated PEG in average size were efficiently separated from 1.5 to 5.6 nm with monodispersity, exhibiting the fine enough degree of density gradient separation (Fig. 5.1a).

Though powerful, aqueous separation of nanoparticles has several limitations: 1. It is only suitable for the aqueous soluble nanoparticles, while a lot of nanoparticles

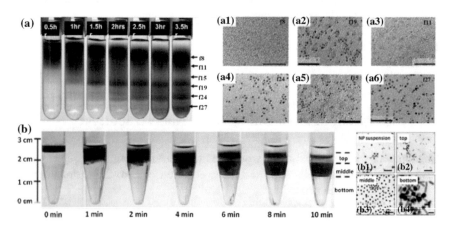

Fig. 5.1 a Optical and TEM images showing the separation of 4 nm FeCo@C nanoparticles (Digital camera images of ultracentrifuge tubes taken at 30 min intervals). (a1–a6) TEM images of different fractions labeled in (**a**). Scale bars: 50 nm; **b** the evolution of the penetration of nanoparticles into an aqueous three-phase system. The solvent of the suspension of nanoparticles (i.e., water) stayed as a clarified layer on top of the system, small nanorods (i.e., the desired product) penetrated slightly into the top phase, small nanospheres migrated to the middle phase, and large particles of both shapes sedimented to the bottom. (b1–b4) TEM images of suspension of nanoparticles (suspension of NP) and samples collected from the layers as shown in (**b**). The scale bar in each image corresponds to 200 nm

were synthesized in organic phase and got solubilized therein (such as Au, CdSe, and Si nanocrystals), and phase transfer might cause serious aggregation. 2. A large mass ratio of salts or solutes are added inside to make the gradient, which significantly complicated the consequent purification procedure to obtain separated nanoparticles. Such separation requirement promotes the emergence of organic phase separation. Therein the gradient media are organic phase (polar or non-polar) to avoid the aggregation of dispersed colloidal nanoparticles, and keep them isolated. Besides, after sampling the product out, the organic media can be vaporized to get "pure" sample. Since unbounded surfactants or soluble by-products could be isolated from the colloidal products, the rate-zonal separation method can also be used for purification of nanocrystals in organic phase.

Additionally, the nanomaterials prepared in organic phase would inevitably aggregate in aqueous solution and should be separated with organic gradient media. Besides, after sampling the product, the residual organic media can be easily removed by various treatments, such as evaporation. Similarly, the simple, rapid, and effective density gradient centrifugation method can also be used for separation and purification of nanocrystals in organic phase. Through dispersing the colloidal nanoparticles in nonhydroxylic solvents using ultracentrifugation in an organic density gradient which gives rapid separation and concomitant purification, Sun's group [4] separated the Au colloidal nanoparticles synthesized in oleylamine. The thin layer (usually 0.1–0.4 mL) of the Au colloidal suspension was placed on a density gradient made by mixing different ratios of cyclohexane and tetrachloromethane (50–90% of CCl_4 by volume; density range, 1.13–1.41 g/cm3). With the centrifugation speed of 50000 rpm ($\sim$330000 g), the Au colloidal particles were finely separated as monodispersed 4.8, 7.2, 8.0, 9.3, and 10.9 nm, and the size of the error range was less than 1.5 nm (Fig. 5.2a). In addition, to make the separation more visible, fluorescent CdSe NPs (CdSe concentration: $\sim$25 mg/mL) were synthesized and separated by using the cyclohexane/tetrachloromethane gradient (Fig. 5.2b).

For the small density of zero-dimensional nanoparticles, if the size is up to several hundred nanometers, such as silica nanoparticles, the rate-zonal centrifugation can be simply used. Chen et al. [5] separated various silica particles with size from 280 nm up to 440 nm successfully in $\sim$1 min by centrifugation. The polydispersity index improved largely, e.g., from 0.116, 0.084, 0.071, and 0.102 to 0.013, 0.007, 0.006, and 0.009 for 80 nm, 260 nm, 740 nm, and 5 lm silica particles, respectively (Fig. 5.3).

While for the very small sized and narrow distributed silicon nanocrystalline, one can change the solvation shell thickness to enlarge the apparent density. Geoffrey A. Ozin et al. [6] modified the silicon nanocrystalline with decylsilane to increase the hydration shell thickness and realized the size separation of the silicon nanocrystalline (Fig. 5.4) by using the 40 wt% 2,4,6-tribromotoluene in chlorobenzene as the density gradient medium. They further investigated the size-dependent structural, optical, electrical, and biological properties of silicon.

This method enables nanoparticles separated in miscible organic solvents. Since a large number of metal and semiconductor colloidal NPs are synthesized in organic

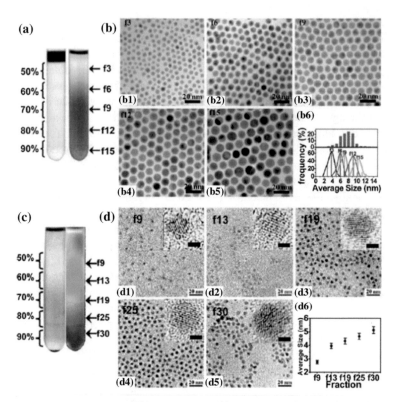

Fig. 5.2 **a** Digital images of ultracentrifuge vessels containing Au nanoparticles before (left vessel) and after (right vessel) separation at 25000 rpm for 12 min. **b** TEM images of typical fractions. The graph in the bottom right corner shows a comparison of the size distribution difference before (red columns in the upper section) and after (colored columns in the lower section) centrifugation separation. Each size histograms was measured from at least 200 particles. **c** Digital camera images of ultracentrifuge vessels containing CdSe nanoparticles using a cyclohexane + tetrachloromethane gradient after separation at 50000 rpm for 60 min. The left image was recorded under white light; the right image was recorded under UV irradiation at 365 nm. **d** HRTEM images of typical CdSe nanoparticle fractions. Magnified individual nanoparticles are shown in the insets (the bars in the insets are 2 nm). The graph in the bottom right corner shows the size evolution of particles along the centrifuge vessel

systems, the separation, purification, and transformation can be employed directly on the raw product mixture by density gradient separation method.

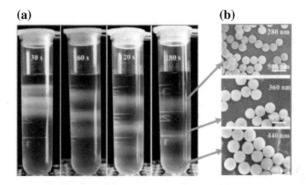

Fig. 5.3 a Centrifugal separation of 280, 360, and 440 nm silica particles, through a gradient of 400, 500, 600, 700, and 800 mg/mL sucrose layered from top to bottom at 25 °C, 600 g, and centrifugal time of 30, 60, 120, and 180 s, respectively. **b** The SEM results illustrate the size and shape

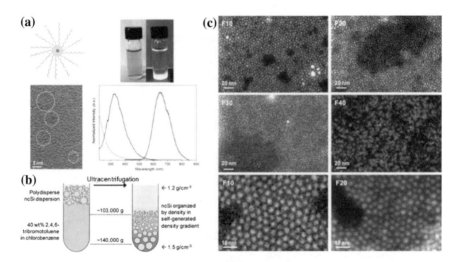

Fig. 5.4 a Decyl-capped silicon nanocrystals; **b** the schematic diagram of the ultracentrifugation process; **c** Size analysis of monodisperse ncSi fractions

5.2   Separation of One-Dimensional Nanostructures

For the separation of one-dimensional nanostructures, the density gradient ultra-centrifugation rate separation (DGURS) method has also usually emerged as an efficient tool for the separation. The kinetic equation of one-dimensional nanos-tructures can be simplified as:

$$\frac{d^2x}{dt^2} + [\frac{3}{2}(r+h)^2(l+2h)]^{-\frac{2}{3}}\frac{9\eta\theta}{2\rho_p}\frac{dx}{dt} + \frac{\rho_m - \rho_p}{\rho_p}\omega^2 x = 0 \qquad (5.2)$$

Similarly, the rate-zonal centrifugation can be simply used for the separation of nanorods with large density (ρ_p). For example, after synthesizing CdS nanorods by solvothermal method with the amine as the solvent, Sun et al. [7] used a five-layer density gradient (30%, 40%, 50%, 60%, and 70% cyclohexane/CCl4 solutions) and centrifugation at 30 000 rpm (relative centrifugal force, RCF = 113600 g) for 70 min. The NRs in the fractions in the bottom half of the tube showed a significant increase in diameter (from ~ 4.2 nm to ~ 6.5 nm) while only a relatively small decrease in length (from ~ 12.3 nm to ~ 10.6 nm). They successfully sorted CdS NRs according to different aspect ratio, and the separated fractions showed significant size-dependent fluorescence property (Fig. 5.5).

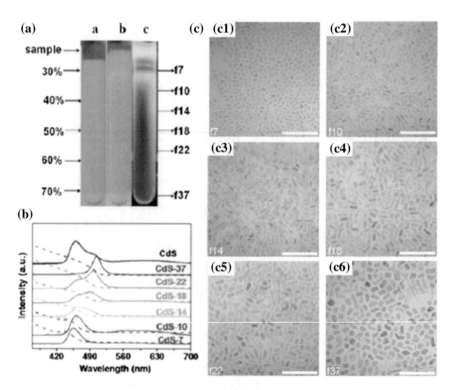

Fig. 5.5 Digital camera images of the ultracentrifuge tubes before and after separation at 30 000 rpm: **a** before separation; after centrifugation for 70 min; separated NRs under 365 nm UV irradiation (fX means the Xth fraction going from top to bottom); **b** Photoluminescence (solid lines) and absorbance spectra (dotted lines) of typical fractions after separation (the black line shows the photoluminescence spectrum of the CdS product before separation); (c1–c6)TEM images of CdS NR fractions separated by DGURS (the scale bar is 50 nm in each case)

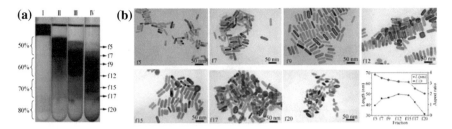

Fig. 5.6 a Digital camera images of the ultracentrifuge tubes before and after separation at 10,000 r/min: (I) before separation; (II) after separation for 10 min; (III) after separation for 20 min; (IV) after separation for 30 min; **b** TEM images of separated Au NRs in typical fractions. The graph in the bottom right corner shows the evolution of the lengths (squares) and aspect ratios (triangles) in the Au NRs in different fractions

Besides, by using different mixing ratios of EG with aqueous CTAB solution (50%–80% of EG by volume), and centrifugation speed with 10,000 r/min (17,000 g) for 10 min, Liu et al. [8] separated the gold nanorods with different length, namely aspect ratios. Au NRs were separated into distinct zones along the centrifuge tube (Fig. 5.6). He et al. [9] also successfully applied density gradient centrifugation method to achieve the separation of gold nanorods (Fig. 5.7).

As for the separation of one-dimensional nanostructures with high density, such as Au nanorods, the sedimentation rate is usually high in ultracentrifugal field. If the target one-dimensional nanostructures are with very similar length, according to the kinetic equation, one can also adjust the viscosity of gradient media to decrease the sedimentation rate and realize efficient separation. By adding the poly (2-ethyl-2-oxazoline), Liu et al. [10] increased the viscosity, changed the sedimentation behaviors of Au nanorods, and successfully separated the Au–NRs with size ranging from 25.6 to 26.1 nm through 5500 g within 5 min, which appears to be the fastest method for separation of Au–NRs (Fig. 5.8).

Different with metal nanostructures, the carbon nanotubes possess extreme small density and diameter ($\sim$0.7–1.1 nm), particularly, the density difference is very small, which remained a big challenge for the effective separation of them. As a breakthrough, Hersam's group [11] first introduced the DGC method to sort carbon nanotubes. Combing with the method of enriching DNA to wrapped semiconducting SWNTs (single-walled carbon nanotubes), they increased the hydration shell thickness and enlarged the density difference. It is worth noting that, even after modification of DNA, the density difference of SWNTs is still very small (1.11–1.17 g cm^{-3}), and SWNTs can be separated by isopycnic mode, with different colored bands. This finding could likely be applied to separate other nanostructures in which external functionalization or relative hydration of the surfaces varies with size (Fig. 5.9).

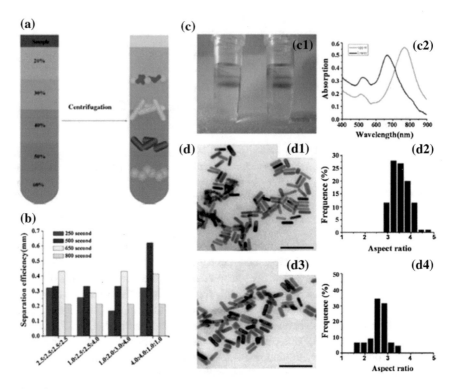

Fig. 5.7 a Predicated separation profile of a complicated sample containing rods with diverse aspect ratios and spheres with different diameters. **b** Calculated separation efficiency of two equal-mass gold nanorods of aspect ratios 2 and 4, respectively, during sucrose gradient centrifugation using different sucrose layer thickness and centrifugal time. **c** Optical image of the centrifuge tube and the UV–visible spectra of the two layers after separation. **d** TEM image and aspect ratio distribution of nanorods from the upper layer, and the lower layer. All particles and rods with aspect ratio less than 1.5 were not counted and did not show up in the histogram data (The scale bar is 100 nm)

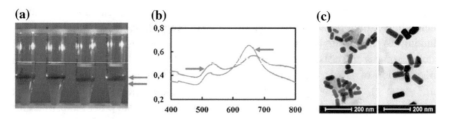

Fig. 5.8 a Pictures of the centrifuge tubes taken at the end of four parallel runs; **b** Normalized UV–VIS spectra of the gold nanorods in the different fractions; **c** TEM images of the gold nanorods in the top and the bottom fractions

Fig. 5.9 Redistribution of iodixanol and SWNTs during ultracentrifugation. **a** Profile of the density gradient before (dotted line) and after (solid line) centrifugation. During centrifugation, the iodixanol redistributed. **b–f** Sedimentation of SWNTs in a density gradient before and after 3.5, 7, 8.75, and 10.5 h of ultracentrifugation

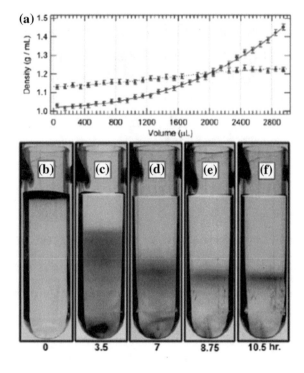

Based on the finding of the isopycnic mode applied on the separation of nanotubes, Green et al. [12] used sodium cholate (SC) and sodium dodecyl sulfate (SDS) rather than DNA which is relatively more prohibitive and cheap to modify the carbon nanotubes, and the purity of the single-walled carbon nanotubes after separation could reach 99% (Fig. 5.10a). Later, Hersam's group [13] used three bile salts (sodium cholate (SC), sodium dodecyl sulfate (SDS), and sodium dodecylbenzene sulfonate (SDBS)) to modify the carbon nanotubes. By the scalable technique of density gradient ultracentrifugation, they isolated narrow distributions of SWNTs in which >97% and within a 0.02nm diameter range (Fig. 5.10b).

Hersam's group [14] has also shown that density gradient ultracentrifugation could be used to separate double-wall nanotubes (DWNTs) from mixtures of single- and multi-wall nanotubes through differences in their density, which could be used in transparent conductors. They added approximately 70% DWNTs into 110 mL of 1 wt% SC aqueous solution, with a loading of ∼2 mg/mL in a steel beaker. Then the CNTs were efficiently sorted from polydispersed mixtures to monodispersed SWNTs with different diameters and DWNTs (Fig. 5.11).

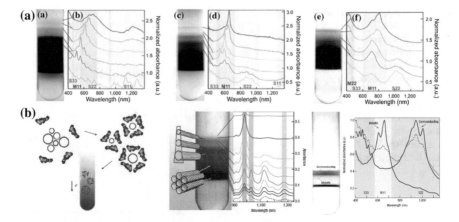

Fig. 5.10 **a** Photographs of ultracentrifuge tubes following DGU, Sorted SWNTs (solid) were collected from gradients at points labeled; **b** Sorting of SWNTs by diameter, bandgap and electronic type using density gradient ultracentrifugation. Schematic of surfactant encapsulation and sorting, where r is density, and Photographs and optical absorbance (1 cm path length) spectra after separation using density gradient ultracentrifugation; Photograph of laser-ablation-grown SWNTs separated

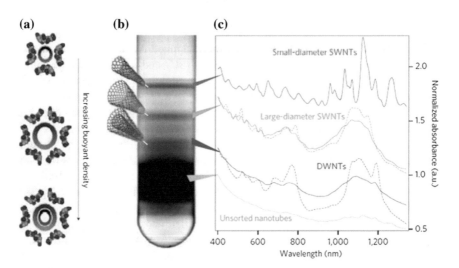

Fig. 5.11 Separation of carbon nanotubes by number of walls using density differentiation. **a** Schematic illustration of carbon nanotube encapsulation by sodium cholate and its effect on nanotube buoyant density. **b** Photograph of a centrifuge tube following the first-iteration density gradient ultracentrifugation (DGU) separation of as-received nanotubes. **c** Optical absorbance spectra of the bands of material taken from the centrifuge tube at the locations indicated

5.3 Separation of Two-Dimensional Nanostructures

Recent years, more and more attentions have been attracted to the 2D materials, such as graphene. The separation of the two-dimensional nanomaterials referred kinetic equation can be simplified as:

$$\frac{d^2x}{dt^2} + [\frac{3}{2}(r+h)^2(l+2h)]^{-\frac{2}{3}}\frac{9\eta\theta}{2\rho_p}\frac{dx}{dt} + \frac{\rho_m - \rho_p}{\rho_p}\omega^2 x = 0 \qquad (5.3)$$

Graphene, as the most widely investigated two-dimensional material, the thickness (r) is very small, inducing the very little density difference for different layered graphene samples. Similarly to the separation of carbon nanotubes, surface modification is an effective choice to increase the hydration shell thickness (t) and enlarge the density difference, and thereby improve the separation efficiency. Sun's group [15, 16] separated the graphene oxide (GO) solution by rate-zonal separation method. Through the typical centrifugation conditions were 5 min at 50,000 rpm ($\sim$300 kg, SW65 Rotor, Beckman Coulter); they finally obtained GO fractions in accordance with the size difference which is found to be related to the oxidation degree (Fig. 5.12b). By studying the properties of separated fractions, it is found that small nanosheets possess weak visible region absorption, strong fluorescence, and high oxidation degree.

The density gradient centrifugation could also be used for the separation of other kinds of two-dimensional materials besides graphene, such as layered double

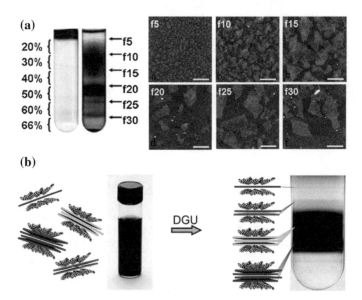

Fig. 5.12 **a** Schematic illustration of the graphene exfoliation process. Graphite flakes are combined with sodium cholate (SC) in aqueous solution; **b** Digital camera images of the ultracentrifuge tubes after separation at 50 K rpm, Tapping-mode AFM images (2 $\times$ 2 μm^2, scale bar: 500 nm) of different fractions of CRG as labeled in left tube

hydroxides and zeolite. Michael Tsapatsis et al. [17] used density gradient centrifugation to purify the zeolite product, through rate-zonal centrifugation of nanosheets in a multilayered density gradient; they separated exfoliated nanosheets from larger unexfoliated particles. The MFI-nanosheets were centrifuged (Beckman Coulter, Avanti J-20 XP equipped with JA25.50 rotor) in four 50 mL FEP centrifuge tubes at 40,000 g for 3 h to sediment zeolite nanosheets at the bottom of the centrifuge tubes. This application example indicated that a nonlinear density gradient of organic solvents could be used to purify exfoliated MFI-nanosheets prepared by melt compounding of multilamellar MFI.

While for the separation of graphene nanosheets with similar size and different layers, even with surface modifications, the density difference is usually still very limited, which requires extremely fine separation. As a breakthrough, Hersam et al. [18] used two-plane amphiphilic surfactant (sodium deoxycholate) to stabilize the package of graphene sheets, and the different thicked fractions can be sorted with controlled thickness by isopycnic separation. During this process, the aqueous solution phase approach is enabled by the planar amphiphilic surfactant sodium cholate (SC), which forms a stable encapsulation layer on each side of the suspended graphene sheets (Fig. 5.12a).

Similarly, by adding cesium chloride (CsCl) to iodixanol, increasing its maximum buoyant density to the point, Hersam et al. [19] sorted the high-density two-dimensional nanomaterial rhenium disulfide (ReS_2) with isopycnic density

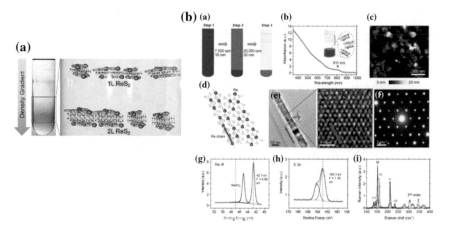

Fig. 5.13 A Schematic illustration of the ReS_2 separated process; **B a** Experimental procedure to enrich few-layer ReS_2 nanosheets. **b** Optical absorbance spectrum of the resulting ReS_2 aqueous dispersion with a peak at 811 nm. Inset: photograph of the as-prepared ReS_2 dispersion with the schematic illustrating ReS_2 nanosheets surrounded by SC. **c** Atomic force microscopy image of solution-processed ReS_2 following deposition on a Si wafer. **d** A schematic of the atomic structure of ReS_2 (blue: Re atom; yellow: S atom; red dotted line: Re chain direction). **e** A transmission electron microscopy image of a ReS_2 nanosheet and a high-resolution TEM image. The red arrow indicates the direction of a Re chain. **f** Selected area electron diffraction pattern of a ReS2 nanosheet. **g, h** X-ray photoelectron spectroscopy data for the (**g**) Re 4f and **h** S 2p core levels. **i** Raman spectrum of ReS_2 nanosheets

gradient ultracentrifugation. After resulting dispersion was centrifuged at 7500 rpm to remove unexfoliated ReS_2 powder, the ReS_2 nanosheets with relatively large lateral sizes were ultracentrifuged at 20 000 rpm to collect (Fig. 5.13). The density gradient ultracentrifugation separation technique has been widely used for the separation of different sizes and dimensions. Even though there is some difference of viscous resistance for different morphologied nanostructures, the apparent density of nanostructures with surface solvation shell is usually the key factor.

5.4 Separation of Assemblies/Clusters

Nanomaterial assemblies have good application prospects in many fields. Compared with monomers, the density of assemblies/cluster is related to the assemble degree and can be considered as different sized single nanoparticles. From a separate point of view, the assemblies/cluster can also be effectively separated. Examples will be given to verify the centrifugation theory (Fig. 5.14).

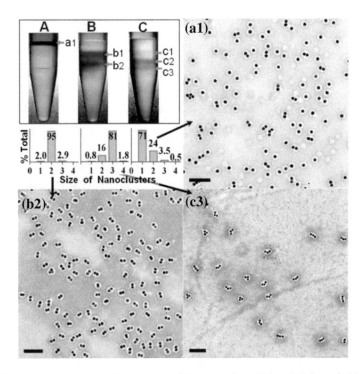

Fig. 5.14 **a** A typical setup of differential centrifugation, where 62% and 11% aq. CsCl and then AuNPn@PSPAA in water were layered from bottom to top. **b** The result of A after 20 min centrifugation. **c** Separation result of a pre-enriched trimer sample. (a1, b2, and c3) TEM images of the respective fractions indicated in **a**–**c** (see large-area views in the Supporting Information); the histograms are shown in the insets. Scale bars: 100 nm

Chen et al. [20, 21] encapsulated gold nanoparticles with amphiphilic diblock copolymers (PSPAA) to form core-shell nanoassemblies. Based on rate-zonal separation, AuNP dimers, and trimers were separated with excellent purity. Since structural intactness is critical for the construction of any nanoassembly, the approach offers a facile separation method without additional inconvenience. Their subsequent research revealed that the structural uniformity of the "hot spots" therein reduced the ambiguities in calculating and interpreting the respective SERS enhancement factors, and the relative intensity ratios of the nanoclusters ($I_{2NP} = 16 I_{1NP}$ and $87 I_{3NP}$) involve few assumptions and are thus more reliable (Figs. 5.15 and 5.16).

Carbon nanodots (CDs) have gained research relevance as novel carbon nano-materials. Due to their extremely high colloidal stability, common ultracentrifu-gation has failed in sorting them. Luo et al. [22] introduce an ultracentrifugation method using a hydrophilicity gradient to sort such non-sedimental CDs. CDs were pre-treated by acetone to form clusters. Such clusters "de-clustered" as they were forced to sediment through media comprising gradients of ethanol and water with varied volume ratios. During the centrifugation, primary CDs with varied sizes and degrees of carbonization detached from the clusters to become well dispersed in the corresponding gradient layers. Their settling level was highly dependent on the varied hydrophilicity and solubility of the environmental media.

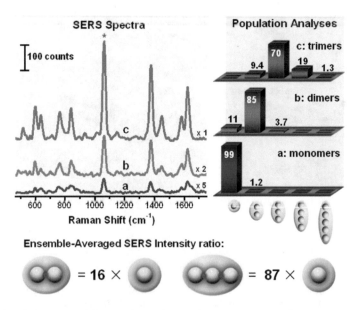

Fig. 5.15 SERS spectra of the samples enriched with **a** monomers, **b** dimers, and **c** trimers of Au@Ag NPs (**d**) 20 nm; excitation: 785 nm at 290 mW; insets: the histograms of these samples). The schematics in the lower panel show the SERS intensity ratio of the nanoclusters

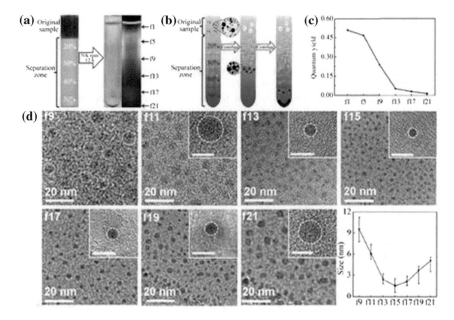

Fig. 5.16 **a** Digital photographs of the centrifuge tube (under ambient light) before and (left: under ambient light, and right: under UV light) after centrifugation. **b** Scheme of proposed mechanism of separation: CDs were clustered at starting point and de-clustered at successive layers with increasing water content during sedimentation. **c** QY variation of typical fractions, showing that more stable CDs have higher fluorescence

References

1. Sun X, Tabakman SM, Seo WS et al (2009) Separation of nanoparticles in a density gradient: FeCo@C and gold nanocrystals. Angew Chem Int Edit 48 (5):939–942
2. Akbulut O, Mace CR, Martinez RV et al (2012) Separation of nanoparticles in aqueous multiphase systems through centrifugation. Nano Lett 12(8):4060–4064
3. Peng W, Mahfouz R, Pan J et al (2013) Gram-scale fractionation of nanodiamonds by density gradient ultracentrifugation. Nanoscale 5(11):5017–5026
4. Bai L, Ma X, Liu J et al (2010) Rapid separation and purification of nanoparticles in organic density gradients. J Am Chem Soc 132(7):2333–2337
5. Hu C, Chen Y (2015) Uniformization of silica particles by theory directed rate-zonal centrifugation to build high quality photonic crystals. Chem Eng J 271:128–134
6. Mastronardi ML, Hennrich F, Henderson EJ et al (2011) Preparation of monodisperse silicon nanocrystals using density gradient ultracentrifugation. J Am Chem Soc 133(31):11928–11931
7. Ma X, Kuang Y, Bai L et al (2011) Experimental and mathematical modeling studies of the separation of zinc blende and wurtzite phases of CdS nanorods by density gradient ultracentrifugation. ACS Nano 5(4):3242–3249
8. Li S, Chang Z, Liu J et al (2011) Separation of gold nanorods using density gradient ultracentrifugation. Nano Res 4(8):723–728
9. Xiong B, Cheng J, Qiao Y et al (2011) Separation of nanorods by density gradient centrifugation. J Chromatogr A 1218(25):3823–3829

10. Dong S, Wang Y, Tu Y et al (2016) Separation of gold nanorods by viscosity gradient centrifugation. Microchim Acta 183(3):1269–1273
11. Arnold MS, Stupp SI, Hersam MC (2005) Enrichment of single-walled carbon nanotubes by diameter in density gradients. Nano Lett 5(4):713–718
12. Green AA, Hersam MC (2008) Colored semitransparent conductive coatings consisting of monodisperse metallic single-walled carbon nanotubes. Nano Lett 8(5):1417–1422
13. Arnold MS, Green AA, Hulvat JF et al (2006) Sorting carbon nanotubes by electronic structure using density differentiation. Nat Nanotechnol 1(1):60–65
14. Green AA, Hersam MC (2009) Processing and properties of highly enriched double-wall carbon nanotubes. Nat Nanotechnol 4(1):64–70
15. Green AA, Hersam MC (2009) Solution phase production of graphene with controlled thickness via density differentiation. Nano Lett 9(12):4031–4036
16. Sun X, Liu Z, Welsher K et al (2008) Nano-graphene oxide for cellular imaging and drug delivery. Nano Res 1(3):203–212
17. Sun X, Luo D, Liu J et al (2010) Monodisperse chemically modified graphene obtained by density gradient ultracentrifugal rate separation. ACS Nano 4(6):3381–3389
18. Agrawal KV, Topuz B, Jiang Z et al (2013) Solution-processable exfoliated zeolite nanosheets purified by density gradient centrifugation. AIChE J 59(9):3458–3467
19. Kang J, Sangwan VK, Wood JD et al (2016) Layer-by-layer sorting of rhenium disulfide via high-density isopycnic density gradient ultracentrifugation. Nano Lett 16(11):7216–7223. https://doi.org/10.1021/acs.nanolett.6b03584
20. Chen G, Wang Y, Tan LH et al (2009) High-purity separation of gold nanoparticle dimers and trimers. J Am Chem Soc 131(12):4218–4219
21. Chen G, Wang Y, Yang M et al (2010) Measuring ensemble-averaged surface-enhanced raman scattering in the hotspots of colloidal nanoparticle dimers and trimers. J Am Chem Soc 132(11):3644–3645
22. Deng L, Wang X, Kuang Y et al (2015) Development of hydrophilicity gradient ultracentrifugation method for photoluminescence investigation of separated non-sedimental carbon dots. Nano Res 8(9):2810–2821

Chapter 6
Application of Nanoseparation in Reaction Mechanism Analysis

Zhao Cai, Xiaohan Qi, Yun Kuang and Qian Zhang

Abstract Density gradient centrifugation has been established to obtain monodisperse nanoparticles with strictly uniform size and morphology, which are usually hard to be obtained by synthetic optimization. Previous chapters have demonstrated the versatility and universality of such separation method, by which nearly all kinds of nanostructures can be separated, including particles, clusters, and assemblies. Further, reaction mechanism, as well as structure–property relationship, can also be investigated based on the separated fractions. The focus of this chapter is the reaction mechanism analysis using density gradient centrifugation, namely by introducing a distinctive functional gradient layer, such as reaction zone and assembly zone, reaction mechanisms can be therefore studied since the reaction time can be pre-designed and the reaction environment can be switched extremely fast in a centrifugal force field. In a word, "lab in a tube" based on nanoseparation opens a new door for the investigation of synthetic optimization, assembly behavior, and surface reaction of various nanostructures.

Keywords Lab in a tube · Reaction zone · Assembly zone · Growth mechanism Ultraconcentration

6.1 The Concept of "lab in a tube"

The concept of "lab in a tube" was first put forward as a micrototal analysis system in the field of biology [1, 2]. In recent decades, in order to extend the functionality of old lab-on-a-chip system [3], lab in a tube has been designed to compress an entire laboratory into a smaller architecture, in which lots of individual detection or analysis components were integrated and each could be used individually or together [1].

In 2012, Sun's group [4] extended this concept to the field of nanoseparation. "lab in a tube" based on nanoseparation is the integration of numerous functional gradient layers into a single centrifuge tube constituting a microsystem of several independent units, each can individually perform its specific role such as "reaction zone" or "assembly zone" (Fig. 6.1).

© The Author(s) 2018
X. Sun et al., *Nanoseparation Using Density Gradient Ultracentrifugation*,
SpringerBriefs in Molecular Science, https://doi.org/10.1007/978-981-10-5190-6_6

Fig. 6.1 Schematic cartoon of "lab in a tube" system based on nanoseparation

Since "lab in a tube" system can combine several functionalities such as separating and assembling nanostructures, it can be applied to various research settings. First of all, by introducing a separation zone, nanoparticles with strictly uniform size can be obtained, and their size–property relationship can be therefore investigated. Second, by analyzing the chemical composition or crystal structure of separated fractions, key pieces of information can be obtained to guide synthetic optimization, which are usually difficult to be got by traditional contrast experiments. Third, by introducing a reaction zone, the corresponding reaction mechanism can be investigated since the reaction time can be finely controlled and the chemical environment can be changed soon. Last but not the least, by introducing an assembly zone, the symmetrical or asymmetric assembly of various nanostructures can also be achieved since their directional movement can be specially designed.

6.2 Size–Property Investigation Through DGUC Nanoseparation

Strictly monodisperse nanoparticles with focused size distribution can be obtained through DGUC nanoseparation, which are usually hard to be got by optimized synthesis and can lay a perfect foundation for their size–property investigation. What is more, even the most advanced synthesis technology cannot meet the high standard for fabrication of monodisperse nanostructures with specific parameters, like ultrashort single-walled carbon nanotubes, DGUC nanoseparation can be used instead to verify theoretical predictions about such nanostructures.

Single-walled carbon nanotubes (SWCNTs) have been extensively studied due to their unique physical and chemical properties, such as the length-dependent optical performance [5–7]. It has been predicted theoretically that the bandgap of a nanotube increases with the length decreasing and quantum confinement effect should be observed as the SWCNTs approach zero-dimensional sizes. However, it is still impossible to obtain monodisperse zero-dimensional SWCNTs by synthetic optimization.

In this case, DGUC nanoseparation can show unique advantages over conventional synthesis methods. As reported by Dai's group in 2008, to obtain ultrashort SWCNTs, low-density gradient layers should be utilized so that the sedimentation velocity of individual SWCNT can be well controlled [8]. As a result, a three-layer step gradient made from 5, 7.5, and 10% iodixanol solutions was used. In addition, the centrifugation time should be accurately controlled at the same time, since insufficient time would lead to an uncompleted separation meanwhile a too-long centrifugation time would cause the sedimentation of all SWCNTs to the bottom of centrifuge tube. After a 3 h centrifugation at a ultrahigh speed (~ 300 kg), SWCNTs 2–50 nm in length could be separated according to their length. As shown in Fig. 6.2, the SWCNTs in fraction 6 were ~ 7.5 nm in length, and the average length of subsequent fractions (f8, f12, and f18) increased gradually to ~ 11, 27, and 58 nm, respectively. Besides, the corresponding length histograms of the fractions further confirmed the successful separation.

The optical characterization results are shown in Fig. 6.3, in which a clearly continuous and monotonic blueshift can be observed for shorter SWCNT fractions. As to the ultrashort SWCNTs fractions, their UV–Vis–NIR absorption and PL peaks blueshifted up to ~ 30 meV compared to that of longer SWCNTs fractions, which should be attributed to quantum confinement effects.

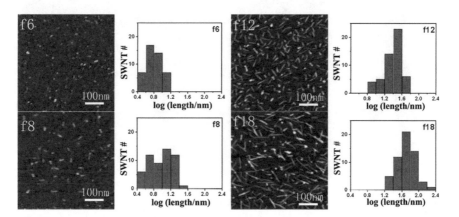

Fig. 6.2 AFM images of SWCNTs in various fractions after separation and the corresponding length histograms of the fractions are shown in the right. Ultrashort SWCNTs (~ 7.5 nm) have been obtained in fraction 6

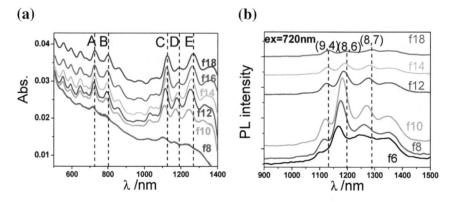

Fig. 6.3 a UV–Vis–NIR absorption spectra and **b** NIR-PL spectrum of SWCNTs in various fractions after separation. Blueshift of the absorption peak could be observed in both spectra for shorter SWCNTs

As another example, cadmium selenide nanoparticles (CdSe NPs) are also well-known for their size-dependent fluorescence; even slight difference in size (i.e. 1 nm) could have a significant impact on their fluorescent properties [9]. The DGUC separation of CdSe NPs can therefore pave a new way for their size–property investigation.

In 2010, Sun's group [10] performed the separation in organic density gradients (cyclohexane + tetrachloromethane) as the CdSe NPs were prepared in organic phase. After DGUC separation at 50000 rpm for 60 min, as shown in Fig. 6.4a, the centrifuge tube under UV irradiation at 365 nm showed different colors of

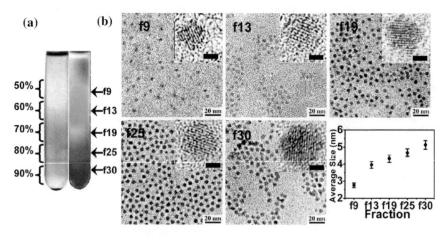

Fig. 6.4 a Digital camera images of ultracentrifuge tubes containing CdSe NPs using an organic gradient after size separation. The left one was recorded under natural light and the right one was recorded under ultraviolet light of 365 nm. **b** HRTEM images of corresponding CdSe particle fractions. The inset presents the magnified representative nanoparticle (the bars: 2 nm) and the size evolution of nanoparticles is shown in the bottom right corner

fluorescence, which gave direct evidence to the size separation since larger nanoparticles with more redshifted emissions could be observed at lower gradient layer. The HRTEM images in Fig. 6.4b further confirmed the successful size separation of CdSe NPs.

As the introduction of polystyrene (PS) into organic gradient layers can significantly increase their viscosity, it should slow down the sedimentation of nanoparticles. As expected, PS-containing gradient (vessel II) showed a limited separation compared to PS-free gradient (vessel I) and only applying longer centrifugation time can the separation be completed (vessel III), and thus a finer separation can be achieved. The fluorescence spectra are shown in Fig. 6.5b and c both demonstrated a clear redshift, suggesting the size evolution. It is also worth noting that the discrimination effect was lost in the last few fractions in PS-free gradients, which in turn indicated a finer separation in PS-containing gradient.

In addition, the introduction of polymer makes it possible to fabricate composite films with the separated CdSe NPs after the volatilization of cyclohexane and tetrachloromethane (Fig. 6.5d). Such highly flexible and transparent films could have great potential applications in various fields, such as labeling and information technology.

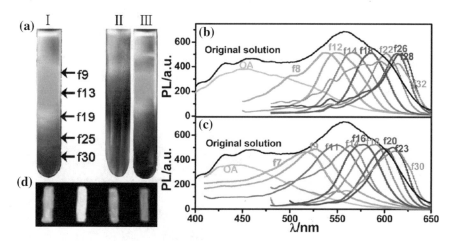

Fig. 6.5 **a** Digital camera images of ultracentrifuge tubes containing CdSe NPs after size separation, recorded under UV irradiation at 365 nm: (vessel I) PS-free gradient, at 50000 rpm for 60 min centrifugation; (vessel II) PS-containing gradient, at 50000 rpm for 60 min centrifugation and (vessel III) PS-containing gradient, at 50000 rpm for 110 min centrifugation. **b** Fluorescence spectra of fractions from vessel I. **c** Fluorescence spectra of fractions from vessel III. **d** Digital camera images of composite strips made from CdSe NPs with different sizes, recorded under UV irradiation at 365 nm (size: 105 mm × 20 mm)

6.3 Synthetic Optimization Through DGUC Nanoseparation

The crystallization behaviors of inorganic nanostructures are often different even in one-pot synthesis since the microenvironment of their nucleation and growth cannot be exactly the same. Synthetic optimization is therefore usually hard to be performed by traditional contrast experiments. In view of the fact that nanoparticles with different size, morphology, or phase can be separated through DGUC nanoseparation, by analyzing the chemical composition or crystal structure of separated fractions, key pieces of information can be obtained and thus guide synthetic optimization [11].

Cadmium sulfide (CdS) semiconducting nanorods (NRs), known as quantum rods, have attracted an enormous amount of attention due to their promising applications. Although extensive effort has been devoted to control the anisotropic growth of CdS NRs, more effective methods should be developed to regulate the nucleation and growth processes to obtain CdS nanorods with precisely tailored electrical and optical properties [12].

Inspired by nanoseparation, in 2011, Sun's group have performed the separation of CdS NRs [13]. After centrifugation in a cyclohexane + tetrachloromethane gradient at 30000 rpm for 135 min, longer CdS NRs could be observed at lower positions of the centrifuge tube, as shown in Fig. 6.6, which suggested that the weight of a NR played the dominant role on its sedimentation rate.

HRTEM images of CdS NR in corresponding fractions in Fig. 6.7a further confirmed the length evolution. Besides, the photoluminescence spectra in Fig. 6.7b showed a blueshifted band-edge emission. The relationship between the length of

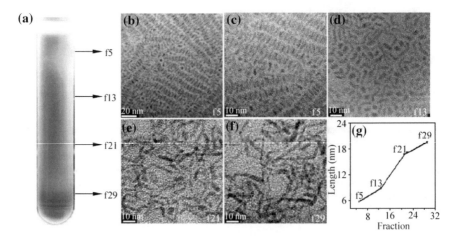

Fig. 6.6 **a** Digital camera images of ultracentrifuge tubes containing CdS NRs after separation, recorded under irradiation at 365 nm. **b–f** TEM images of corresponding CdS NR fractions. **g** Length evolution of CdS NRs in different fractions

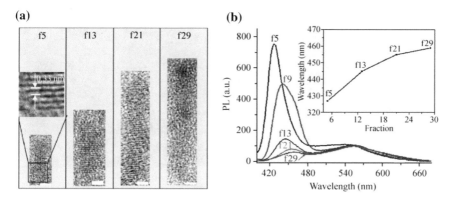

Fig. 6.7 **a** Typical HRTEM images and **b** photoluminescence spectra of CdS NRs fractions (scale bar: 2 nm, the inset presents a plot of the band-edge emission peak positions of corresponding fractions)

CdS NRs and their photoluminescence properties became apparent after separation, which inspired them to selectively prepared monodisperse samples by synthetic optimization.

As reported by Papadimitrakopoulos, the presence of oxygen can promote the transformation of CdSe quantum dots to nanorods. Here, CdS nanorods were synthesized in N_2, air, and O_2 atmosphere, respectively. As expected, N–CdS contained the shortest nanorods with photoluminescence dominated by short wavelength band-edge emission while O–CdS NRs were the longest and long wavelength surface-trap emission predominated (Fig. 6.7). This indicated that oxygen-deficient condition could lead to the formation of ultrashort CdS nanorods, which were nearly the same as that obtained in f5 after separation of A–CdS (Fig. 6.8).

In 2013, Sun's group also studied the phase transition of Yb^{3+} and Er^{3+} co-doped $NaYF_4$ nanocrystals ($NaYF_4$:Yb^{3+}/Er^{3+} NCs) using density gradient ultracentrifuge separation [14]. The $NaYF_4$:Yb^{3+}/Er^{3+} NCs were synthesized in an oleic acid–water–ethanol system via hydrothermal process. After separation, the fraction 1 was orange colored under excitation at 980 nm, and the middle fraction 6 turned red while fraction 18 at the bottom exhibited green emissions, as shown in Fig. 6.9.

The structural and compositional difference among these three fractions was investigated using transmission electron microscopy. f1 and f14 were composed of ∼28 and ∼44 nm nanocubes, respectively, meanwhile, f18 was composed of ∼1.2 um nanorods. It was found that f1 and f14 had relatively high Y atomic ratio and f18 was richer in Yb and Er.

The size, morphology and composition difference inspired Sun's group to verify the sharp and composition evolution by monitoring the formation process by TEM and EDS at different reaction time points (Fig. 6.10). When the reaction lasted for

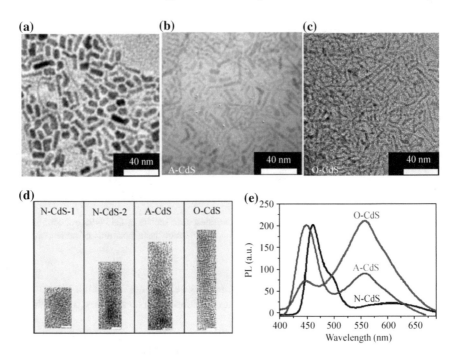

Fig. 6.8 a–c TEM images of CdS NRs prepared in N₂ (N–CdS), air (A–CdS), and O₂ (O–CdS), respectively. **d** Typical HRTEM images and **e** photoluminescence spectra of corresponding CdS NRs (there are two kinds of rods in N–CdS, which are denoted as "N-CdS-1" and "N-CdS-2")

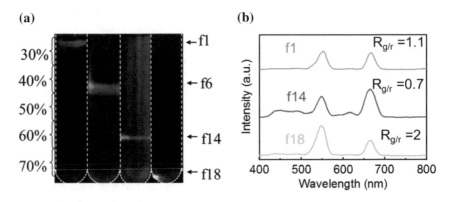

Fig. 6.9 a Digital camera images of ultracentrifuge tubes containing NaYF₄:Yb³⁺/Er³⁺ NCs after separation under excitation at 980 nm; **b** UC fluorescence spectra of typical NaYF₄:Yb³⁺/Er³⁺ NCs fractions

1.5 h, pure α-phase NaYF₄:Yb³⁺/Er³⁺ nanocubes were obtained and the Y:Yb:Er atomic ratio was found to be 83:16:1, which was nearly the same as f1. As the reaction time was prolonged to 20 h, mixed phases were appeared, indicating a

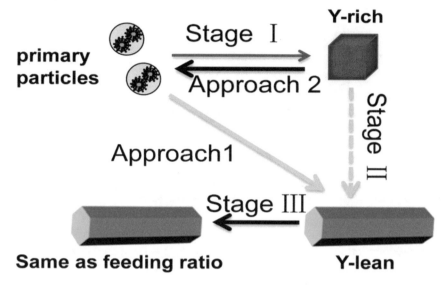

Fig. 6.10 Schematic illustration of the formation process of NaYF$_4$:Yb^{3+}/Er^{3+} NCs

phase transition. At last, when the reaction lasted for 7 days, the atomic ratio of Y: Yb:Er returned to 78:20:2, which was the same as the feeding ratio. Thus, a plausible mechanism of morphological and compositional evolution was proposed. Namely, small cubes (f1) with a higher Y content were formed at the initial stage, the subsequent phase transition (stage II) led to the formation of thermally stable β-phase nanorods, which were rich in Yb. After the dissolution–crystallization equilibrium (stage III), the Y:Yb:Er atomic ratio turned the same as the feeding ratio, and the nanorods became bigger in size.

6.4   Surface Reaction Mechanism Investigation Through "Reaction Zone" in the Density Gradient

To date, the investigation on surface reaction mechanism still critically relies on the capture of reaction intermediates [15–17]. However, high surface area of NPs endows NPs with high reactivity, which lead to a quick reaction rate, but unfortunately, such intermediates are usually hard to be obtained in a short time through traditional centrifugation methods.

DGUC nanoseparation method provides new opportunity of isolating intermediate NPs within a short period of time. By introducing a reaction zone in gradient layers, surface reaction mechanism can be investigated, since the reaction time can be finely controlled and the chemical environment can be changed very soon.

As an example, to investigate the surface reaction mechanism of galvanic replacement reaction between Au and Ag, a 20% to 70% EG/H$_2$O gradient layers

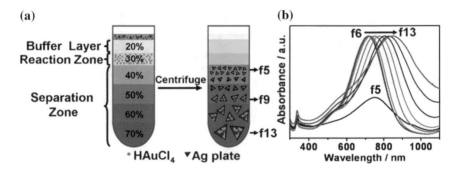

Fig. 6.11 a Schematic illustration of surface reaction mechanism investigation through DGUC nanoseparation by introducing a reaction zone. **b** UV–Vis spectroscopy of the corresponding fractions

were used in centrifugation [4]. As shown in Fig. 6.11 a, the second layer was set as reaction zone by introducing a certain amount of reactant "HAuCl$_4$" and the first layer was used as a buffer layer to prevent a direct mixing and reaction. Furthermore, the lower four layers acted as the separation zone to separate the reacted Ag nanoplates by their sizes. Ag nanoplates prepared in aqueous phase was placed on the top of the gradient layers. As shown in Fig. 6.11 b, after separation, a clearly redshifting could be observed, demonstrating the increased size of reacted Ag nanoplates from f6 to f13. However, f5 did not show the similar trend, which should be attributed to the hollow structures.

The hollow structure of Ag nanoplates in f5 was further confirmed by TEM, as shown in Fig. 6.11a. Ag nanoplates with smaller size should have slower sedimentation rate, which could lead to a longer exposure time and thus result in their hollow structures. On the contrary, bigger Ag nanoplates should have a shorter exposure time, which should result in a short-time reaction. As estimated, the exposure time of f9 in reaction zone was only 30 ± 17 s, which is much shorter than that can be achieved by traditional reaction and centrifugal process.

To get deeper insight into the surface reaction mechanism, the Ag nanoplates in f9 were characterized by HRTEM and EDS. The Au/Ag atomic ratio of the edge regions was measured to be 0.231, much higher than other regions, indicating the edge side should be the favored site for such surface reaction. Besides, the Au/Ag atomic ratio of thick and thin part of the basal plane was 0.064 and 0, respectively. This means the thin part presents the Ag dissolution zone and the thick part should be responsible for the Au deposition.

On the basis of the results above, the structural evolution of triangular Ag nanoplates during the surface reaction can be divided into two stages, as shown in Fig. 6.12c. At the initial stage, the reaction starts at the edge side and at the same time on the basal plane surface. With the reaction time increasing, Ag dissolution and Au deposition jointly lead to the formation of hollow structure when the reaction comes to its end at stage II.

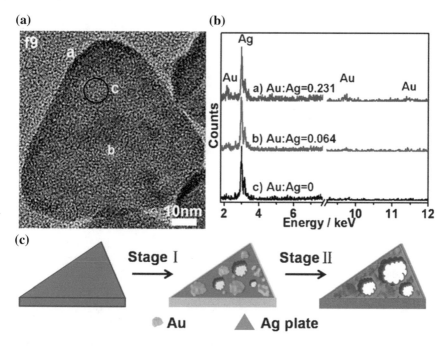

Fig. 6.12 **A** HRTEM image of typical reacted Ag nanoplate in f9 and **B** the EDS results of corresponding regions as marked in the HRTEM image: **a** edge; **b** thick part and **c** thin part of the basal plane. **C** Schematic illustration of the structural evolution of a triangular Ag nanoplate during the surface reaction

6.5   Controlled Asymmetric Assembly Through "Assembly Zone" in the Density Gradient

Controlled assembly of NPs is critical for the investigation on their collective properties, which is of great importance in guiding the fabrication of elaborate nanodevices [18, 19]. However, currently, random Brownian motions remained the only way to achieve the asymmetric assembly of NPs and such uncontrollable method has greatly limited the application of the assembly structures. Since centrifugal field can be applied to overcome the Brownian motion effect of NPs, DGUC separation can be designed to make colloidal heteroassembly by introducing an "assembly zone" in the density gradient layers [20].

During the centrifugal process, the directional motion of bigger NPs should be faster than that of NPs with smaller size and thus symmetric heteroassemblies can be fabricated by a "crash reaction" as schematically shown in Fig. 6.13a. Big Au NPs, ∼60 nm in diameter and with a positively charged surface, were placed on the top of the density gradient layers. Meanwhile, ∼20 nm Au NPs with a negatively charged surface were set at a lower layer, with a buffer layer inserted to avoid a spontaneous assembly. When a large centrifugal force was applied, big Au NPs

(a)   **(b)**

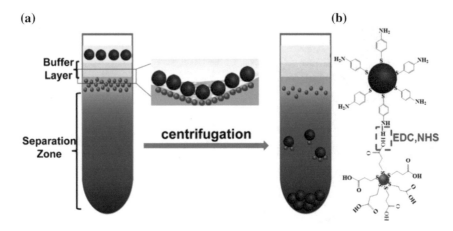

Fig. 6.13 **a** Schematic illustration of controlled asymmetric assembly of different sized Au NPs through "crash reaction" in density gradient centrifugation. **b** Schematic linkage of functionalized Au NPs through dehydration condensation reaction

can cross through the buffer layer and react with small ones with opposite charge (Fig. 6.13b) [21].

Figure 6.14 shows a typical result of asymmetric assembly in the density gradient. After the "crash reaction," the UV–Vis spectra showed a slight redshift from f18 to f22 (Fig. 6.14b), demonstrating the successful assembly of Au NPs. Besides, TEM images in Fig. 6.14c, d further confirmed the asymmetric assembly. However, it should be noted that the assembling efficiency was not high enough, resulting in only a small portion of effective collision.

6.6   Ultraconcentration of Colloidal NPs Through Water/ Oil Interfaces

As a high-efficiency separation and purification way density gradient centrifugation can avoid colloidal nanoparticles from nanostructure destruction and aggregation. This separation is appropriate to aqueous phase or organic phase. However, what will happen as the nanostructures passing through oil–water interface in the centrifuge tube during centrifugation? Kuang et al. had a detailed study in this aspect. Density gradient centrifugation of colloidal NPs through water/oil interface becomes a high-efficiency way to achieve NPs purification and concentration without any aggregation [22]. The interface between different layers of density gradient concentrate the nanoparticles while separation progress. NPs will pass through the interface after the enrichment (Fig. 6.15a). For certain nanoparticles, small droplet could be formed under high centrifugal force which leads fast

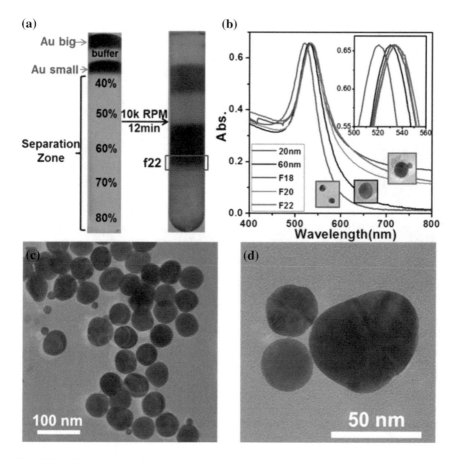

Fig. 6.14 a Digital camera images of the density gradient layers before and after "crash reaction" in centrifugation. **b** UV–Vis spectra of corresponding fractions after separation. **c** and **d** TEM images of the Au NP assemblies in f22

sedimentation, while larger droplet formed under low centrifugal force which needs more time to accumulate at interface.

Ultraconcentration of colloidal NPs through water/oil interfaces apply to zero-dimensional nanoparticles, one-dimensional nanomaterials, and two-dimensional nanosheets. While some low-density nanosheets can't pass through water/oil interface, such as graphene and graphene oxide because centrifugal force is not enough to overcome buoyancy force and interfacial tension, thus graphene oxide will concentrate at the interface.

The volume of bottom ultraconcentration colloidal NPs was too small to measure, spectroscopy linear curve fitting method was used to test volume and concentration while diluting. In order to ensure the measurement is accurate, it is necessary to use organic liquid seal keep the ultraconcentration volume same. In order to make sure that density of 30% CCl_4/cyclohexane liquid seal is lower than

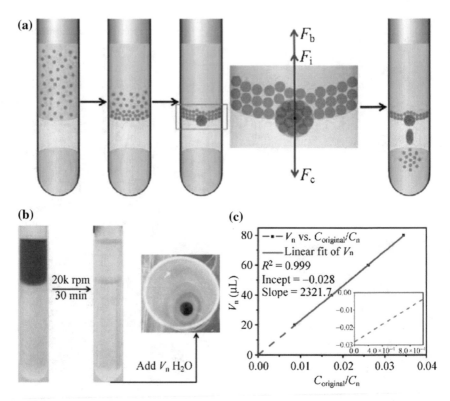

Fig. 6.15 **a** Schematic illustration of droplet sedimentation. **b** Gradient centrifugation in centrifuge tube with minimized the volume of bottom layer. **c** Graph between V_n and $C_{original}/C_n$

water, add cyclohexane to the organic layer so the water won't float on the organic layer when diluting the ultraconcentration colloidal NPs.

$$C_n(V_n + V_0) = C_0 V_0$$

Volume and concentration of ultraconcentration colloidal NPs assume as V_0 and C_0, respectively. Add volume water V_n to dilute the colloidal solution concentration to C_n. Based on the Beer–Lambert law, C_n could be calculated by UV–Vis spectroscopy. Liquid seal has ensured the V_0 stay the same during the dilution. In this premise, the concentration of original Au nanoparticle assumes as $C_{original}$ that the formula transformed into $V_n = C_0 V_0/C_n - V_0 = (C_0/C_{original}) V_0/(C_n/C_{original}) - V_0$. Therefore, after calculation of the $C_0/C_{original}$ by Beer–Lambert law, a linear plot with intercept $-V_0$ and slope equal to $(C_0/C_{original}) V_0$ can be obtained. By fitting the slope and intercept, ultraconcentration solution volume and relative increased multiples to the original solution can be calculated.

After calculation, 3 ml Au colloidal NPs pass through water/oil interface was concentrated to 0.028 μl and the concentration increased at least 10^4 times. In such high concentration, the density of ultraconcentration solution up to 5.28 g/ml and the packing density of Au NPs up to 64.2% which is close to the limit of closely packed packing density (74%). Herein, the ultraconcentration Au colloidal NPs were staying in the closely packed state which almost presented a solid state. Under such high packing density, the colloid NPs still without aggregation and maximize the dispersed state. Such a high-efficient concentration proves that the water/oil interface centrifugation method is very efficient on purification. The research mentioned above 64.2% volume of ultraconcentration colloid was occupied by Au NPs. So there was only 0.01 μl original solution pass through the water/oil interface which means the purification way could remove the 99.99% impurity one-time only. Such high-efficient way demonstrates superiority than other separation methods.

References

1. Duan R, Zuo X, Wang S et al (2013) Lab in a tube: ultrasensitive detection of microRNAs at the single-cell level and in breast cancer patients using quadratic isothermal amplification. J Am Chem Soc 135(12):4604–4607
2. Smith EJ, Schulze S, Kiravittaya S et al (2010) Lab-in-a-tube: detection of individual mouse cells for analysis in flexible split-wall microtube resonator sensors. Nano Lett 11(10):4037–4042
3. Harazim SM, Quiñones VAB, Kiravittaya S et al (2012) Lab-in-a-tube: on-chip integration of glass optofluidic ring resonators for label-free sensing applications. Lab Chip 12(15):2649–2655
4. Zhang C, Luo L, Luo J et al (2012) A process-analysis microsystem based on density gradient centrifugation and its application in the study of the galvanic replacement mechanism of Ag nanoplates with HAuCl4. Chem Commun 48(58):7241–7243
5. Liu C, Fan Y, Liu M et al (1999) Hydrogen storage in single-walled carbon nanotubes at room temperature. Science 286(5442):1127–1129
6. Odom TW, Huang J-L, Kim P et al (1998) Atomic structure and electronic properties of single-walled carbon nanotubes. Nature 391(6662):62–64
7. Bachilo SM, Strano MS, Kittrell C et al (2002) Structure-assigned optical spectra of single-walled carbon nanotubes. Science 298(5602):2361–2366
8. Sun X, Zaric S, Daranciang D et al (2008) Optical properties of ultrashort semiconducting single-walled carbon nanotube capsules down to sub-10 nm. J Am Chem Soc 130(20):6551–6555
9. Murray C, Norris DJ, Bawendi MG (1993) Synthesis and characterization of nearly monodisperse CdE (E = sulfur, selenium, tellurium) semiconductor nanocrystallites. J Am Chem Soc 115(19):8706–8715
10. Bai L, Ma X, Liu J, Sun X et al (2010) Rapid separation and purification of nanoparticles in organic density gradients. J Am Chem Soc 132(7):2333–2337
11. Ma X, Kuang Y, Bai L et al (2011) Experimental and mathematical modeling studies of the separation of zinc blende and wurtzite phases of CdS nanorods by density gradient ultracentrifugation. ACS Nano 5(4):3242–3249
12. Zhang G, He P, Ma X et al (2012) Understanding the "Tailoring Synthesis" of CdS nanorods by O₂. Inorg Chem 51(3):1302–1308
13. Sun X, Ma X, Bai L et al (2011) Nanoseparation-inspired manipulation of the synthesis of CdS nanorods. Nano Res 4(2):226–232

14. Song S, Kuang Y, Liu J et al (2013) Separation and phase transition investigation of Yb 3 +/ Er 3 + co-doped NaYF4 nanoparticles. Dalton T 42(37):13315–13318
15. Chatterjee D, Deutschmann O, Warnatz J (2002) Detailed surface reaction mechanism in a three-way catalyst. Faraday Discuss 119:371–384
16. Long R, Yang R (2002) Reaction mechanism of selective catalytic reduction of NO with NH3 over Fe–ZSM-5 catalyst. J Catal 207(2):224–231
17. Koop J, Deutschmann O (2009) Detailed surface reaction mechanism for Pt-catalyzed abatement of automotive exhaust gases. Appl Catal B Environ 91(1):47–58
18. Chen G, Wang Y, Yang M et al (2010) Measuring ensemble-averaged surface-enhanced Raman scattering in the hotspots of colloidal nanoparticle dimers and trimers. J Am Chem Soc 132(11):3644–3645
19. Urban AS, Shen X, Wang Y et al (2013) Three-dimensional plasmonic nanoclusters. Nano Lett 13(9):4399–4403
20. Qi X, Li M, Kuang Y et al (2015) Controllable assembly and separation of colloidal nanoparticles through a one-tube synthesis based on density gradient centrifugation. Chem Eur J 21(19):7211–7216
21. Song S, Kuang Y, Luo L et al (2014) Asymmetric hetero-assembly of colloidal nanoparticles through "crash reaction" in a centrifugal field. Dalton Trans 43(16):5994–5997
22. Kuang Y, Song S, Liu X et al (2014) Solvent switching and purification of colloidal nanoparticles through water/oil Interfaces within a density gradient. Nano Res 7(11):1670–1679

Printed in the United States
By Bookmasters